T0361750

SACRED
SCIENCE

SACRED
UNDERSTANDING DIVINE CREATION
SCIENCE

William H. West, M.D.

Post Hill
PRESS

A POST HILL PRESS BOOK
ISBN: 979-8-88845-888-4
ISBN (eBook): 979-8-88845-889-1

Sacred Science:
Understanding Divine Creation
© 2025 by William H. West, M.D.
All Rights Reserved

Cover design by Jim Villaflores

Post Hill Press
New York • Nashville
posthillpress.com

Published in the United States of America
1 2 3 4 5 6 7 8 9 10

CONTENTS

Section Three: Consciousness

Supplements

OVERVIEW

For much of modern history, science and faith have shared an essential worldview: God created the heavens, the Earth, and every creature to live on the Earth. He also made men and women in His image. Many of the greatest scientific minds through the centuries have been devoutly religious. You know their last names: Copernicus, Galileo, Newton, Kepler, Descartes, Pascal, and many others, and that only takes us through the 1700s. The most influential physicists of the nineteenth century—Faraday and Maxwell—were also devout people of faith. In their minds, the pursuit of science involved reading the thoughts of God.[1]

In the late seventeenth century, Francis Bacon defined the scientific method required to decipher God's creative tools, using inductive reasoning and careful observation. Bacon saw science as a divine gift and considered scientific experimentation a means of glorifying God.[2]

---

[1]   Max Caspar, *Kepler* (New York: Dover Publications, 1993), 62.
[2]   Stanford Encyclopedia of Philosophy, s.v. "Francis Bacon," by Jürgen Klein, Winter 2016 ed., https://plato.stanford.edu/archives/win2016/entries/francis-bacon/.

Isaac Newton was a leading physicist, mathematician, philosopher astronomer and theologian, widely recognized as one of the greatest geniuses to ever live, matched only by Einstein. He invented calculus and formulated the first theory of gravity. Newton was deeply religious, writing extensively about prophecy and the history of early religion and the Christian church.

In 1687, Isaac Newton used that scientific method to decipher one of nature's greatest secrets, becoming the first great scientist to express his work in mathematical terms through his equations for gravity. Consolidating his reputation as a natural philosopher of enormous stature, Newton declared himself a servant of God, commissioned to reveal truths of nature to humankind. He saw God as a divine mechanic.[3]

---

[3]   James Gleick, *Isaac Newton* (New York: Vintage Books, 2003), 110.

In the same century, Nicolaus Copernicus developed a sun-centered cosmos. Many scientists of the day considered God the ultimate geometer.

These founders of modern science created an entirely new way of knowing God, referred to as "natural theology," a branch of theological argument that celebrates the created order as divine.[4] Science and religion were allies.

In 1859, Charles Darwin, an English geologist, biologist, and naturalist, published *On the Origin of Species*, dismissing God's supervision of life.[5] In the Darwinian worldview, evolution had no purpose or direction. Species are a product of random variation amplified by natural selection. Humans were a happy accident and nothing more. The science/faith alliance absorbed a potentially fatal blow.

The Rise of Scientism

Darwin cast doubt on the concept of a creator; his acolytes have grown bolder over time. A radical form of materialism has carried the materialist mantra to the extreme. Science, the claim is made, is the only source of truth. In his book *Scientism: Philosophy and the Infatuation with Science*, the philosopher Tom Sorell describes scientism as follows: "Scientism is a matter of putting too high a value on natural science in comparison with other branches of learning or culture."[6]

MIT physicist Ian Hutchinson has a similar but slightly more extreme definition: "Science, modelled on the natural sciences, is

4   N. T. Wright, *History and Eschatology: Jesus and the Promise of Natural Theology* (Waco: Baylor University Press, 2018).
5   Charles Darwin, *On the Origin of the Species* (London: John Murray, 1859).
6   Tom Sorell, *Scientism: Philosophy and the Infatuation with Science* (London: Routledge, 1991).

the only source of real knowledge."[7] Science, in this worldview, is the sole judge of truth. Given the startling success of science in the past few decades—from the personal computer and smartphones to the medications and vaccines that protect our health—many people feel that they face a bimodal choice. Either they believe in science as their primary source of truth, or they stick with the faith of their forefathers. "Scientism" pushes even harder, dismissing any non-scientific proposal, including religious belief, as idle speculation. This corrosive message is impacting church attendance and public surveys

In 1975, for example, at least 75 percent of Americans belonged to a religious congregation. By 2020, that percentage had fallen below 50 percent.[8] One in three Americans now declares their religion as "none." So do 44 percent of people aged eighteen to thirty.[9] The percentage of Americans who do not attend religious services has increased from 9 percent to 30 percent.[10] These data reveal a massive shift in a short period of time

One physicist and spokesperson for scientism explained his worldview: "I respect the notions of God and other divine beings. However, I insist on one thing. I insist that any statements made by such beings about the material world, including statements recorded in the sacred books, must be subject to the experimental

---

7   Ian Hutchinson, *Monopolizing Knowledge: A Scientist Refutes Religion-Denying, Reason-Destroying Scientism* (Belmont: Fias Publishing, 2011), vii.

8   Bob Smietana, "Gallup: Number of Americans who Belong to a Church or House of Worship Plummets," Religion News Service, April 18, 2019, https://religionnews.com/2019/04/18/gallup-number-of-americans-who-belong-to-a-church-or-house-of-worship-plummets/.

9   Timothy Beal, "Can Religion Still Speak to Younger Americans?" *Wall Street Journal,* November 14, 2019, https://www.wsj.com/articles/can-religion-still-speak-to-younger-americans-11573747161.

10   Emma Green, "It's Hard to Go to Church," *The Atlantic,* August 23, 2016, https://www.theatlantic.com/politics/archive/2016/08/religious-participation-survey/496940/.

testing of science. In my view, the truth of such statements cannot be assumed. They must be tested and revised or rejected as needed."[11]

This author combines his dismissive attitude toward scripture with a depressing worldview: "The universe is made of material and nothing more,...[it is] governed exclusively by a small number of fundamental forces and laws."[12] So much for music, art, or Shakespeare, much less Holy Scripture.

A prophet of scientism thus delivers an ancient challenge to the Creator of the universe: "Show us a sign." God himself should be subject "to the experimental testing of science." There is a delicious irony to this distorted worldview: *Everything that exists is a sign.* Divine algorithms display their work across the heavens. The entire universe demonstrates God's creative powers.

Real Science

As with any philosophy taken to extremes, "scientism" is a failed worldview. But science unburdened by philosophical nonsense is a majestic endeavor. Science involves "[t]he systematic study of the structure and behavior of the physical and natural world through observation and experiment."[13] The following features are fundamental to the conduct of science:[14]

1. Science is "empirical." Scientists make progress through observation, not logic or philosophy.

---

[11] Alan Lightman, "Meditations on Fact and Faith," *Commercial Appeal*, April 11, 2018, https://www.commercialappeal.com/story/opinion/contributors/2018/04/11/opinion-meditations-science-and-religion-fact-and-faith/504956002/.

[12] Lightman, "Meditations on Fact and Faith."

[13] "Science and Technology," Oxford Reference, accessed July 24, 2024, https://www.oxfordreference.com/page/134.

[14] James C. Zimring, *What Science Is and How It Really Works* (Cambridge: Cambridge University Press, 2019).

2. Good science is analytical, critical, systematic, and, most importantly, reproducible. Different scientists can test a hypothesis and obtain comparable results.

3. Vitally important, science is falsifiable.

Valid theories stand the test of time. They can be put to the test and withstand the challenge. This feature has posed a near-insurmountable (and probably fatal) blow to string theory, the hypothesis that the universe has hidden higher dimensions. No one has defined a test to prove string theory's validity, much less its falsifiability.[15] The concept of a multiverse—a proposal that an infinite succession of universes explains the fine-tuning of our own—fails the falsifiability test as well.

Focused on their area of individual expertise, most scientists avoid a God discussion. They may be searching for gravitational waves or a quantum theory of gravity but give little thought to gravity as a major force of creation. They may be planning the architecture of a quantum computer but ignore the vital role of countless quantum events in the construction of the universe.

But wittingly or not, scientists have collectively produced a remarkable story that drives a stunning conclusion: eight blueprints underlie the unfolding of our universe. In their individual and collective action, they represent undeniable divine design. Equally stunning blueprints underlie life and mind. The conclusion: mind, not mindless matter, has willed us into being.

---

[15] Sabine Hossenfelder, *Lost in Math: How Beauty Leads Physics Astray* (New York: Basic Books, 2018).

How To Use This Book

I have made every effort to make this book accessible to individuals with minimal prior exposure to science, providing a description of blueprints before each section and sprinkling "Key Definitions" throughout the chapters. This will hopefully make the material more understandable for individual readers and those who approach *Sacred Science* as a group.

Many of the most important supporting documents for the book, particularly those involving life and DNA, have been published in just the past few years. Even a key blueprint of the universe, the Higgs boson that encodes the universe with mass, was only confirmed in 2012. Those of us who completed schooling decades ago will be encountering key elements of creation for the first time. Ours is a privileged generation: science has sufficiently matured to supplement scripture as a means of acknowledging and worshipping our Creator.

The book is divided into three sections. The first reveals the eight essential blueprints that printed out our cosmos and created planet Earth. Each one bristles with divine implications. The only explanation for the entire ensemble: God is real and the reason we exist.

Section 2 presents a new theory for macroevolution that updates the standard theory. Many of us have one serious problem with Darwin: by emphasizing randomness and chance, he denied God a role in the process. As people of faith should have suspected all along, God superintended life's journey. This section will outline the components of His plan.

Finally, section 3 confronts issues that science is ill-equipped to address. How does our material brain support higher consciousness?

What do experiences with near death and psychedelics tell us about the primacy of love and the reality of the Holy Spirit?

Additional material is provided in a section labeled "Supplements." The supplements are incidental to the book's major conclusions, but kudos to those who take the time to read them.

Additional Suggestions

I am aware of two groups who might approach this book with apprehension.

The first are those individuals steeped in science. You are already familiar with quarks and Higgs bosons and know that gravity is a product of space-time. You consider science an independent source of knowledge with no relationship to a faith-based worldview. So, then read the conclusion to section 1 and answer this question: How many times can we attribute near-miraculous blueprints/equations of the universe to the accidental work of Mother Nature? These eight interactive blueprints work as a team to produce staggering complexity and beauty. They only make sense as the product of an unfathomable mind who anticipated and deployed their creative power.

Read chapter 17 to learn about gene expression in the lungfish, a "living fossil" with the largest genome of any species. The lungfish fin contains the gene for future toes (*hoxd13*), but only in half of its fin. *Hoxd13* provides notice of a "coming attraction"—a promissory note that digits will replace the fin, ultimately sending life to dry land. But *hoxd13* in half of a fin provides no immediate advantage to the fish. How do those findings fit with "natural selection" as the guiding force for the creation of new species?

Then finish your reading with chapter 26. Respected neuroscientists report that six out of ten atheists treated with a single dose of the psychedelic psilocybin believe they have encountered a loving consciousness that people of faith might consider the Holy Spirit. As a result, they have abandoned the atheist label.

A second group deserves independent consideration. Their question: Why do I care about quarks or a Higgs mechanism that encoded those quarks with mass? What difference does it make to me that God preloaded the genome of a fish with the requirements for living on land?

That group might complete the same reading and ponder the following questions: How many of us walk on a "rare"[16] and "privileged"[17] planet, yet never celebrate its origin and fitness? How many of us experience compassion and love, yet never ponder their source?

This book is written with one hope in mind: that more of us see science in a different light, not as a competitor to scripture-based faith, but as a complementary revelation of God's "eternal power and divine nature" (Romans 1:20, New International Version).

Many people of faith declare their belief by reciting the Apostles' Creed, beginning with the following sentence: "I believe in God the Father, Maker of the Heavens and the Earth."[18] How ironic that those same people of faith distrust the scientific product of their Maker. Fifty percent of Christians dismiss the reality of the big bang, despite its description in the first few sentences of the Bible. Atheistic Darwinism is responsible for this confusion—if science wants to remove God from the unfolding of life, why should we trust any science?

---

[16] Peter D. Ward and Donald Brownlee, *Rare Earth: Why Complex Life is Uncommon in the Universe* (New York: Copernicus, 2000).
[17] Guillermo Gonzalez and Jay W. Richards, *The Privileged Planet: How Our Place in the Cosmos Is Designed for Discovery* (Washington DC: Regnery Publishing, 2004).
[18] en.wikipedia.org/wiki/Apostles-Creed

There is only one way to place science and faith in a proper perspective—review the blueprints of creation enumerated in each section. Upon reading each one, ask the same question: did nature come up with this creative impulse on her own? Then put those blueprints in their holistic perspective and reach an unavoidable conclusion: their symphonic collaboration reflects timeless design and intention.

God asks that we love Him with our heart, soul, and strength—and, yes, with our minds. That includes understanding and embracing sacred science.

Note to Theologians

This book is focused on recent science as a revelation of God. Although it is filled with action verbs ("to create," "to establish libraries in a higher dimension," "to commission and command," etcetera), the author recognizes that the nature of God is beyond human understanding.

Some readers will reject any flavor of "evolution" or a creation process extending over billions of years while others will be comfortable with both concepts. The author has a simple resolution: as a component of your faith in God, remember He makes all things possible.

One final point: while the author approaches this discussion of science and faith with a firm Christian perspective and a strong belief in the triune God and the divinity of Christ, it is hoped that readers of all faiths and even those with no faith at all achieve new insight into the miraculous nature of our universe, our life, and the greatest miracle of all: the ability of our minds to decipher and understand sacred science.

INTRODUCTION

Almost three thousand years ago, King David gazed at the nighttime sky and offered great praise: "Lord, our Lord, how majestic is your name in all the earth! You have set your glory in the heavens."[19]

Humanity has always been fascinated by the structure of the universe and the miracles and mysteries of life, no matter the decade, the location, or the culture. So, it is not surprising that at many points in our lives, most human beings face recurrent fundamental questions, such as:

1. Who or what produced this wondrous universe?
2. Why am I here?
3. Does my life have a purpose?

Beneath these vital questions lie the most important questions: Is there a personal God?

4. Do I have a soul?
5. What happens when I die?
6. Will I ever see my loved ones, parents, or deceased friends again?
7. Will my children, grandchildren, and even my beloved pets be with me after death?

Various theologians, philosophers, scientists, and others of some renown have suggested the answers to these questions. Those answers seem to fall into the following two categories.

---

[19]   Psalm 8:1 (New International Version)

WILLIAM H. WEST, M.D.

Answer One: The Universe is Matter in Motion with No Supernatural Input.

The world is material through and through, a spontaneous eruption from a mysterious abyss. We are the product of an impersonal nature, the product of unguided evolution with no particular purpose or reason. In this model, materialists see God as nonexistent or part of the universe itself—just fields and forces floating in the cosmos.

Should this answer include a mention of "God," He is shoehorned into two models:

1. Pantheism

God is everything, and everything is in God. Albert Einstein embraced this model of God, rejecting a personal God for a non-personal organizing force "who reveals himself in the orderly harmony of what exists, not in a God who concerns himself with fates and actions of human beings."[20]

2. Deism

God set the universe into motion and departed the scene, which means no God at all. Okay, so where did He go?

Philosophers consider this worldview to be naturalism and summarize it as follows.

---

[20] Einstein, Albert (11 October 2010). Calaprice, Alice (ed.). The Ultimate Quotable Einstein. Princeton University Press. p. 325.

> **Key Definition: Naturalism**
> *Natural laws and forces created and control the universe, leaving no room for supernatural or spiritual explanations.*

Advocates of these worldviews acknowledge the mystery of this universe and the appearance of design and order. Still, they attribute this natural order to "matter in motion."

This worldview has gathered steam from several angles, including the battle over biological evolution, evil as manifested in totalitarian governments and terrorist acts, and the artificial concept of conflict between science and faith.

Proponents of answer one often promote the baseless concept that science and faith are engaged in a never-ending war. They point to neo-Darwinian evolution as their primary case in point.[21] Our existence, they declare, is a random accident. Human beings are the chance product of minor variations conspiring with the blind force of nature. Rerun the tape of life and *Homo sapiens* will never show up in the sequel.

Darwinism has been a loud voice magnifying the material impulses of our culture.

But what if neo-Darwinism proves to be fundamentally ill-conceived? What if we learn that Darwinism provides an excellent account of microevolution but falls far short when it comes to the creation of complex new species and genetic adaptations? These

---

[21]   Richard Dawkins, *The God Delusion* (Boston: Mariner Books, 2008).

secular doubters are missing it, big time. More thoughtful observers are reaching this conclusion,[22] which brings us to answer two.

Answer Two: The Universe, Earth, and Life are Products of Divine Creation.

In this answer, a personal God created the universe, shaping its features with human beings in mind, even creating men and women in His image. This God is intensely concerned with the "doings of mankind." While outside of the material world and unencumbered by time, He dispatches His spirit to engage with the minds of men and women, inspiring scripture and promoting love and compassion.

While prioritizing the gift of free will, the God of theism occasionally interferes with the created order to accomplish His ultimate will. He even allows evil to exist for a while to achieve a more significant purpose—humans with a tested character who value His love.

Philosophers summarize this worldview as follows:

Key Definition: Theism

The belief that a supreme being created the universe and continues to supervise its unfolding. Judaism, Christianity, and Islam consider this being to be God.

---

22 David Gelernter, "Giving Up Darwin: A Fond Farewell to a Brilliant and Beautiful Theory," *Claremont Review of Books*, Spring 2019, https://claremontreviewofbooks.com/giving-up-darwin/.

Implications of Mathematics

Mathematical equations share attributes of the divine. They are eternally true, "ideal forms" that describe every component of creation.

Scientist marvel at the unreasonable effectiveness of mathematics in the natural science. Mathematics carries intimations of the divine. As journalist Jerry Bowyer summarizes New Testament and biblical interpretation professor Vern Poythress's argument in his book, *Chance and the Sovereignty of God*: mathematics has "attributes which indicate an origin in God. They are true everywhere (omnipresent), true always (eternal), cannot be defied or defeated (omnipotent), and are rational and have language characteristics (which makes them personal). Omnipresent, omnipotent, eternal, personal… Sounds like God."[23]

---

[23] Jerry Bowyer, "God in Mathematics," *Forbes*, April 19, 2016, https://www.forbes.com/sites/jerrybowyer/2016/04/19/where-does-math-come-from-a-mathematiciantheologian-talks-about-the-limits-of-numbers/.

Every Feature of Our Universe Has a Mathematical Foundation

An unimaginable mathematician and creator planned our universe before the first beam of light, creating a vast library of mathematical blueprints capable of describing every object in the created order. In a step that defies human understanding, He "breathed fire" into a selection of those blueprints and dispatched them on a mission: interact with each other in a creative ballet and build a universe as interesting as possible.

We will visit those essential blueprints in this book's first section. We will analyze their individual contributions. More importantly, we will behold their collective product.

Let There be Light

Holy Scripture gave us the first taste of this mind-bending creation story: "The world was formless and void, and God said: Let there be light."[24] But the theistic worldview has given us much more than a creation story. It has given us a foundation for ethical behavior, including the Ten Commandments. It has provided a foundation for Western civilization that places a high value on education, medical care, art, culture, and philosophy. This theistic worldview created the preconditions for capitalism, including the rule of law and a mechanism for resolving disputes, accompanied by a hunger for discovery and an entrepreneurial spirit.

---

[24]   Genesis 1:3 (NIV).

Is It Realistic to See Science As Complimentary to Scripture?

The organization of science can frustrate the layperson's effort to absorb the larger picture. The mention of quarks and bosons, black holes and singularities, mutations and neural networks may send many of us running for cover. How is it possible to blend that intimidating body of knowledge with a faith-based understanding of creation?

Our first thought is to proceed with great caution. But the time has arrived for caution to take a back seat.

With a little effort, we can know as much about cosmology as Albert Einstein. We can possess a modern understanding of life that would flummox Charles Darwin. We can encounter the world of neuroscience in ways Sigmund Freud could only imagine. Cosmology, evolutionary biology, and neuroscience tell a fantastic story with an identical conclusion: God planned every step of creation before the first flash of light.

Before we examine the blueprints commissioned to create the universe, this book should be placed in further context. Few topics have engendered more debate than the relationship between science and faith. How do arguments presented in the following chapters relate to the controversies that have stirred this heated field for decades, in particular intelligent design[25] and a multiverse?[26]

Intelligent design advocates have attempted to refute random Darwinism with an essential hypothesis: components of life are irreducibly complex, containing multiple interactive parts that

---

[25] Thomas Y. Lo et al., *Evolution and Intelligent Design in a Nutshell* (Seattle: Discovery Institute Press, 2020).
[26] Brian Greene, *The Hidden Reality: Parallel Universes and the Deep Laws of the Cosmos* (New York: Vintage Books, 2011).

wouldn't function if one of the parts was missing. How, the argument goes, would an irreducibly complex system emerge through successive small modifications from less complex systems when an incomplete system brought no advantage to the cell? Their favorite example is the flagellum, the appendage protruding from a wide range of organisms, including bacteria and sperm cells.

The word "flagellum" derives from the Latin word for a whip. Its function requires a rotary motor and a universal joint constructed from over twenty-five proteins. But why would natural selection preserve nonfunctional parts or proteins before the flagellum served a useful function?

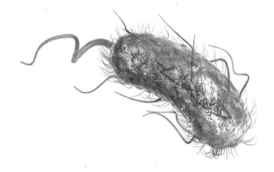

Bacterium with flagella. The Intelligent Design movement has argued there was no reason for random evolution to preserve a partially formed, nonfunctional flagellum.

Why would a step-by-step process of unguided evolution preserve the parts of this essential appendage until the entire apparatus was complete?[27] The flagellum must be the product of an intelligent mind.

---

[27]   Michael J. Behe, *Darwin's Black Box: The Biochemical Challenge to Evolution* (New York: Free Press, 2006).

Neo-Darwinists were quick to dismiss this argument. They posit that components of the flagellum served other needs of the cell before they came together to form a functional rotor. The flagellum might be an evolved extension of an older and simpler device used to expel degraded proteins from the cell.[28]

In the following pages, we will learn about eight blueprints that make the universe a star-printing machine. Delete any essential blueprint, and the universe disappears. Our universe, to use familiar language, is irreducibly complex.

Here's the bad news for those who would dismiss these interactive impulses as accidental products of "nature" randomly commingling over time: these blueprints were present at creation. Interacting to perfection at their cosmic birth, there was no need to refine them postpartum.

But wait, you say, there are infinite universes, a cascading multiverse. Our irreducibly complex components must be a dowry from an unknown and unknowable past.

The Existence of a "Multiverse" to Explain the Life-Friendly Features of Our World

Skeptics use a hypothetical multiverse to escape the implications of fine-tuning in the only universe we know.[29] No scientist denies the fine-tuned features of our universe. The scientific literature and well-respected books have cataloged the wide-ranging extent

---

[28]   Bailey Milne-Davies, Stephan Wimmi, and Andreas Diepold, "Adaptivity and Dynamics in Type III Secretion Systems," *Molecular Microbiology* 115, no. 3 (2021): 395–411, https://doi.org/10.1111/mmi.14658.

[29]   Leonard Susskind, *The Cosmic Landscape: String Theory and the Illusion of Intelligent Design* (New York: Back Bay Books, 2005).

of fine-tuning in the universe,[30] from the mass of the electron to the fact that ice floats, the universe is life-friendly in multiple ways. But there is no need, atheists declare, to see life-friendly tuning as a reflection of a creator.

All things are possible in an infinite succession of randomizing worlds. To the skeptic, an untestable hypothesis of many worlds is preferable to the explanatory power and potential judgment of an all-knowing god.

A divine creator may have chosen to create a multitude of worlds—who are we to restrict His creative options? But He did leave His mark on *this* universe, which makes the multiverse irrelevant. Not only did God commission blueprints for the universe at large, but he also left an additional blueprint on a rare and privileged Earth: He deployed the information molecule called DNA and used its multi-layered codes to create a rising tide of intelligence. He created one special species that can understand and celebrate His science.

Mind Created Matter to Reproduce Mind

It is time to reinvigorate the concept of "natural theology."[31] Holy Scripture, the book of God's word, gives us a fundamental portrait of God. But an understanding of God's scientific works provides a powerful and complimentary approach to understanding God. A step-by-step tour of the cosmos in section 1 provides a striking case in point. The printing machine called the universe was no

---

30   Martin Rees, *Just Six Numbers: The Deep Forces that Shape the Universe* (New York: Basic Books, 2000); Geraint F. Lewis and Luke A. Barens, *A Fortunate Universe: Life in a Finely Tuned Cosmos* (Cambridge: Cambridge University Press, 2016).
31   William Lane Craig and J. P. Moreland, eds., *The Blackwell Companion to Natural Theology* (Hoboken: Wiley Blackwell, 2009).

accidental product of nature. God designed each one of its parts to accomplish His ultimate will: there would be an expanding universe and a just-right sun nurturing a life-supporting planet called Earth.

Be prepared to understand a significant take-home point from the universe as a harbinger of our study of life. God dispatched blueprints for the cosmos from a vast library containing all known mathematical equations. That library includes the description of every potential component of creation. Mathematicians have explored its stacks to retrieve descriptors for electricity, gravity, asteroids, comets, orbits, and quasars. They have even explored that library to solve the mysteries of black holes.[32] God's library is exhaustive, and His blueprints are eternal.

In section 2, we will apply the lessons of the cosmos to the unfolding of the species and reach an even more remarkable conclusion: God also built a library for life. From a bacterium to a blue whale—and yes, to human beings—sequences for life's structures (combinations of letters of DNA that constitute a gene), like the blueprints of the universe, are stored in an eternal library.[33] The science of the cosmos is sacred. So are the DNA sequences of life.

Section 3 affirms God's ultimate intention. Nor only did His DNA build the most complex material object in the universe, the human brain, but he also made it an instrument for pursuing and understanding His science.

John Polkinghorne, an English theoretical physicist, theologian, and Anglican priest, was a prominent voice explaining the relationship between science and religion. A professor of mathematical physics at the University of Cambridge, he encouraged all

---

[32] Roger Penrose, *The Road to Reality: A Complete Guide to the Laws of the Universe* (New York: Vintage Books, 2004), 691.

[33] Andreas Wagner, *Arrival of the Fittest: Solving Evolution's Greatest Puzzle* (New York: Current, 2014).

who would listen to consider science and religion reciprocal views of reality. Each worldview informs and enriches the other: "If people in this so-called 'scientific age' knew a bit more about science than many of them actually do, they'd find it easier to share my view."[34] That view: *"science and faith are complementary paths to a fuller understanding of God."*

This book should be dedicated to Polkinghorne's advice. Learn a little bit more science, dear readers, and encounter the reality of God. Let science and faith give you "binocular vision" as you celebrate God's creative power.

Automobile Metaphor

Like a modern automobile, essential interactive components created the universe. Visualize the universe as a Rolls Royce of cosmic design.

Throughout section 1, you will find automobile metaphors at the ends of the chapters. Some readers may feel that the science of creation is too challenging to understand. Yet, without being a

---

34   John Polkinghorne, *Quarks, Chaos and Christianity: Questions to Science and Religion* (Chestnut Ridge, Crossroad, 1994), xii.

certified mechanic, most drivers are familiar with their cars. They fill up the gas tank and use the engine's power to accelerate down the road. That's a fitting metaphor for the big bang that filled the universe with more energy than we can imagine and continues to drive cosmic expansion.

Most drivers realize they depend on a carburetor to convert gasoline into flammable vapor. That's an appropriate metaphor for the triple-team action that converts tiny particles called quarks into energy-rich protons ready to light up the stars. Chapter 2 describes a unique field that encodes quarks with mass. Chapter 3 describes the code for three-quark teams and a strong nuclear force that ties the proton package together. Both cars and the stars need their fuel.

How much time did it take for the newborn universe to create enough hydrogen gas (protons) to fuel a billion trillion stars including our life-supporting sun? Twenty minutes.

Drivers know they depend on batteries and generators to maintain electric power. It's no surprise, then, that electrons and the electromagnetic force join atomic nuclei produced by nuclear fusion in the stars to build atoms, molecules, and the chemistry of life.

Finally, drivers depend on smooth pavement and good tires and shocks to smooth out the quality of their ride. Space-time and its gravity should feel like an old friend, then, as Earth enjoys a smooth ride across space-time's superhighways.

These metaphors will only take us so far. Imagine your car arriving in an explosion of light and gradually self-assembling in your driveway.

Rather than being intimidated by scientific jargon, approach the scientific blueprints of creation as components of a classic car. In the following chapters, you will meet our Rolls-Royce universe and its essential parts. Some of them, such as "energy," "gravity," and "electricity" are familiar terms. Other notions, like the teams of tiny quarks that build the protons and neutrons of the atom, the Higgs mechanism that encodes those quarks with mass, or the strong and weak nuclear forces that stabilize atoms and trigger the shining of the sun, may be unfamiliar. Great scientists have spent their entire careers exploring these blueprints in detail, often beginning with their mathematical foundation. They gave us a priceless gift. We can ignore math, grasp the big picture, and enjoy one aha moment after another as we recognize the fundamental nature of their discoveries: they have been exploring the divine foundation of creation.

SECTION ONE
The Universe

THE BLUEPRINTS OF THE UNIVERSE

This is a simplified introduction to the blueprints commissioned to create the universe. You will likely recognize a few of the terms from popular science, ideas that will be covered in depth in this section.

$E = mc^2$

Energy and matter are different forms of the same thing, like water, ice, and steam. This equivalence between energy and matter allowed much of the energy of the big bang to be placed in long-term storage in an abundant form of matter, the protons of hydrogen gas. Trillions of future stars have tapped that energy warehouse, using protons to fuel their nuclear fusion.

The energy for *all future stars* was stored away in the *first twenty minutes of creation*. Our sun consumes six hundred billion tons of those hydrogen protons per second. The creation of protons from tiny particles called quarks required two creative steps involving three additional blueprints outlined below.

Quark Code

Quarks build the protons and neutrons for the nucleus of the atom by joining in teams of three. Two up quarks provide two-thirds positive charge each, for a total of four-thirds positive charge. One down quark provides a one-third negative charge. The net charge of the proton? Positive one, matching the minus-one charge of the electron. With only one up quark but two down, the net charge of the neutron is a convenient zero. Neutrons play a critical role during construction of heavier elements in the stars.

The Higgs Field and Boson

Quarks and their fellow particles emerged from big bang energy traveling at the speed of light. If they continued to travel at light speed, quarks would have never joined in three-quark teams to form protons or neutrons. But quarks instantly encountered a mass-encoding field, the Higgs field, the only such field known to science. Forced to swim through this mass-encoding ocean, quarks slowed down, took on mass, and began to interact with each other.

Gluons of the Strong Nuclear Force

Like furiously vibrating rubber bands, gluons lock quarks in the confines of the proton and neutron. Gluons assure the stability of the atom and play a vital role during nuclear fusion in the stars.

Space-Time

Like multi-layered bubble wrap, space-time provided a container for the contents of the big bang and the gravitational power to build galaxies and stars. Gravity conspired with the next two blueprints to trigger nuclear fusion in the first ancient star (W particles of the weak nuclear force and the quantum wave function that define and facilitate particle interaction are discussed below).

W Particles of the Weak Nuclear Force

In the hot and dense center of a soon-to-be star, gluons of the strong force hold two protons close together (since protons both carry a positive charge, they repel each other). Then W particles of another nuclear force, the weak force, plunge into a up quark in one of the two protons, converting an up quark to a down. Two protons instantly became a proton-neutron form of hydrogen called deuterium. By repeating this process, the weak force creates a second form of hydrogen called tritium (one proton, two neutrons). Deuterium and tritium then fuse into helium, releasing a small percentage of the starting matter as energy—the heat and light of the stars.

The Quantum Wave Function

By allowing fundamental particles such as protons and neutrons to exist in a superposition, existing at multiple locations in the same instant, and to entangle (join forces), quantum properties allow deuterium and tritium to fuse, creating helium in the process and releasing a small amount of matter as pure energy, as previously described. Nuclear forces and quantum properties conspire with the squeeze of gravity to create the grandeur of the nighttime sky. All particles and their potential are described by a quantum wave function.

The Electromagnetic Force

As nuclear fusion in the stars produced a table of elements and supernova explosions produced by the death of those stars discharged those elements into interstellar space, electrons joined with the nuclei of elements to produce complete atoms. By sharing electrons, atoms create larger chemicals called molecules (for instance, two hydrogen atoms join a single oxygen atom to create H2O, or water).

As the blueprints of the universe created trillions of stars and filled the stretches of space-time with elements, atoms, and molecules, the laboratory of life began to open on planet Earth.

CHAPTER ONE

Let There Be Light

"In the beginning, God created the heavens and the Earth."[35] In a dramatic first step, He released vast quantities of heat and light from a point smaller than an atom. The commandment "Let there be light" launched an expanding universe virtually infinite in size today.

The eruption of the universe in a blinding burst of light is described by Holy Scripture in the first few sentences of Genesis. Science refers to this incredible moment as the "big bang."[36] Time and space had a beginning in infinite heat and light. In the NIV, Genesis 1:1 declared, "Let there be light," and big bang energy delivered more than light. It also gave birth to particles and forces capable of weaving the world. This *creatio ex nihilo* represents a confirmation of faith and a barrier to the reach of science.

The big bang delivered the light of a billion, trillion suns. That quantity of energy and light defies human understanding. But even the vast light of creation would eventually grow dim, barely detectable today in the darkness of space ("the cosmic background radiation"). How would God assure that His commandment "Let there be light" lights up His entire creation?

---

[35] Genesis 1:2–3 (NIV).
[36] Paul Halpern, *Flashes of Creation: George Gamow, Fred Hoyle, and the Great Big Bang Debate* (New York: Basic Books, 2021), 17.

The solution is breathtaking: God commissioned blueprints embedded in the light, each based on advanced mathematical equations, to create a warehouse of stored energy that would someday light up the stars.

Albert Einstein considered the leading scientific thinker of the modern era. He helped launch the field of quantum mechanics with Max Planck and won the Nobel Prize for his work with the photoelectric effect. Einstein is best known for his explanation of gravity as the warping of space-time and the famous equation E=mc2 recognizing that energy and matter are different forms of the same thing.

How could God create a warehouse of stored energy? What would that warehouse accomplish? Einstein discovered the answer to those questions in the most important equation known to science:[37] $E = mc^2$.

---

[37] Albert Einstein, "Does the Inertia of a Body Depend Upon Its Energy Content?" *Annelen der Physik*, September 27, 1905.

Matter can store vast quantities of energy! The speed of light squared is a number equaling 299,792,458 meters per second.[38]

In less than twenty minutes, blueprints of creation filled an energy warehouse destined to fuel trillions of stars. To restate: in less than half an hour, divine blueprints created the raw material for a glorious cosmos filled with galaxies and stars.

The Equivalence of Energy and Matter

Produced today in diverse forms including solar, wind, and chemical, electromagnetic energy launched the creation of our universe with unimaginable quantities of heat and light. That energy had no weight and occupied no space. In contrast, matter consists of objects that have mass and occupy space, ranging from atoms to planets and people. Here's a shocking fact: energy and matter are different forms of the same thing. They are equivalent, like steam from the kettle, flowing water, and frozen ice. As hard as it is to imagine, Earth and its contents, including every atom in our body, began their journey in light.

Einstein's famous equation, $E = mc^2$, describes God's energy plan.[39] Big bang light stored unimaginable quantities of energy (E) in the protons of hydrogen gas (m), creating massive warehouses of potential energy that would someday fuel trillions of stars. (Remember: c^2 is an eighteen-digit number reflecting the enormous capacity of matter to store energy.)

A deeper analysis reveals the divine elegance of creation. Protons and neutrons, particles that create the nucleus of the atom,

---

[38] Jean-Philippe Uzan, "The Fundamental Constants and Their Variation: Observational Status and Theoretical Motivations," *Reviews of Modern Physics* 75, no. 2 (April–June 2003): 403, https://doi.org/10.1103/RevModPhys.75.403.

[39] David Bodanis, *E=mc²: A Biography of the World's Most Famous Equation* (New York: Berkley Publishing Group, 2000).

are derivative products. They carry three tiny passengers called quarks.[40] In the newborn universe, quarks were created in unimaginable numbers by the collision of high-energy photons of light. But before quarks could become matter and join in three-quark teams to create protons and neutrons, they had to be

1. first, encoded with mass by a one-of-a-kind Higgs field (chapter 2)[41] and
2. then packed in teams of three—two up quarks, one down—and locked into place by furiously vibrating rubber bands (technically, "gluons" of a strong nuclear force (chapter 3).[42]

Let there be quarks, encode those quarks with mass, glue quarks into hydrogen protons, and light up the universe with stars.

Question:
Is it likely that an accidental "burp" in the cosmic emptiness delivered a perfect combination of interactive parts working as a creative team?

---

[40] M. Y. Han, *Quarks and Gluons: A Century of Particle Charges* (Singapore: World Scientific Publishing, 1999).
[41] Peter W. Higgs, "Broken Symmetries and the Masses of Gauge Bosons," *Physics Review Letters* 13, no. 16 (October 1964): 508, https://doi.org/10.1103/PhysRevLett.13.508.
[42] Rolf Ent, Thomas Ullrich, and Raju Venugopalan, "The Glue That Binds Us," *Scientific American Magazine* 312, no.5 (May 2015): 42, https://doi.org/10.1038/scientificamerican0515-42.

Loving God with All of Our Mind

Holy Scripture gives us our most intimate portrait of the power and priorities of God. His number one commandment: that we love Him with our heart, soul, strength, and mind. The inclusion of mind carries special meaning for our science-savvy world. God has given us a front-row seat and powerful tools to comprehend His creative powers. As a component of loving God, we should use our God-given minds to understand and celebrate His science.

For many of us, that may seem easier said than done. No matter the quality of someone's education, few have had meaningful exposure to cosmology, particle physics, stellar chemistry, or planetary science. Our high school (or college) courses have hardly brought us up to speed. Why should we care? Because an understanding of the creative forces that have sculpted our universe drives home a powerful message: God is real. He created our universe in an act of divine love. We can acknowledge that love by understanding and celebrating creation.

We can begin that celebration by understanding the most significant equation in science ($E = mc^2$). This beautiful equation allows light to congeal into matter (objects with mass that occupy space). Unfortunately, humankind has also learned to run the equation in reverse, allowing matter to release the destructive power of its stored energy,[43] a fact only too familiar in a world awash in nuclear weapons.

---

[43]   Cynthia C. Kelly, ed., *The Manhattan Project: The Birth of the Atomic Bomb in the Words of Its Creators, Eyewitnesses, and Historians* (New York: Black Dog and Leventhal Publishers, 2020).

> **Key Definition: Energy**
> *The ability to perform work or cause change. Sources of energy include the sun and wind, chemical reactions, electric current, and the mechanical properties of motion. Unlike matter, energy has no mass and occupies no space.*
>
> **Key Definition: Matter**
> *A physical substance containing mass that occupies space. Matter includes atoms, molecules, and their combinations including stars, planets, and people.*

The equivalence of energy and matter is a vital blueprint of creation. Big bang light would fade over time, but stars would defeat the darkness. Einstein's famous equation $E = mc^2$ describes God's plan for the stars.

Evidence for the Big Bang

There is plenty of good science to support the creation of the universe in a big bang. For the initial twenty minutes of creation, the newborn universe was so intensely hot, it functioned as a nuclear fusion reactor, creating hydrogen and smaller amounts of other light elements, including helium and lithium.

Scientists have calculated the percentage of light elements in the universe today: hydrogen is at 75 percent and helium 24 percent, with trace quantities of lithium. How do those quantities jibe with the big bang theory? They are right on the money, suggesting that we can trust the big bang model.[44]

Scientists also see imprints of the big bang in the "cosmic microwave background" or CMB, faint waves of light still detectable as static on our radios and TVs.[45] Scientists consider the CMB the afterglow of the big bang.

Couple the CMB and the proportion of light elements with well-documented evidence that our universe is constantly expanding, and we find science and Holy Scripture in agreement: our universe erupted from nothingness in response to a divine command!

Einstein's Contribution to Big Bang Theory

Albert Einstein is best known for the general theory of relativity, published in 1915, explaining gravity as the warping of space-time. But his "miracle year" happened a decade earlier, in 1905. Working in obscurity at a patent office in Switzerland, Einstein published four back-to-back articles that changed the world of physics. The most famous of those papers established the equivalence of energy and matter ($E = mc^2$).[46] Matter is the storage form of energy. Like water and ice, energy and matter are different forms of the same thing.

---

[44] Alain Coc and Elisabeth Vangioni, "Primordial Nucleosynthesis," *International Journal of Modern Physics E* 26, no. 8 (August 2017): Article 1741002, https://doi.org/10.1142/S0218301317410026.

[45] Erik M. Leitch, "What is the Cosmic Microwave Background Radiation?" *Scientific American*, November 1, 2004, https://www.scientificamerican.com/article/what-is-the-cosmic-microw/.

[46] Bodanis, *E=mc²*.

Blueprints in the Light

The take-home message of this chapter and the two chapters that follow should be repeated, hoping that these once unfamiliar terms grow familiar in the process:

1. The big bang delivered more energy than a million trillion suns.
2. The equation $E = mc^2$ allowed light to congeal into matter.
3. The most abundant matter in the early universe were up and down quarks that create the protons and neutrons of the atom.
4. Initially, massless quarks were moving through the universe at light speed. A Higgs field encoded those quarks with mass, slowing them down and allowing them to interact with each other (chapter 2).
5. A "code of the quarks" divided quarks into three-quark teams that gave protons and neutrons just-right quantities of electric charge. One team (two up quarks, one down) create protons. A second team (one up quark, two down) create neutrons (chapter 3).
6. "Gluons" of the strong nuclear force locked those quarks in place, preparing protons and neutrons for a long, stable ride to the first star (chapter 3).

Protons of hydrogen gas flowed into the space-time container, destined to create deuterium and tritium, the fuel for the first ancient star. Subsequent stars would produce a precious fruit: a table of elements for planets and life. God's commandment "Let there be light" was satisfied trillions of times over as stars filled the expanding cosmos.

Celebrating God's Initial Equation

In this first chapter, we have met four of the eight blueprints commissioned to create the universe: $E = mc^2$, the Higgs field, the quark code, and the strong nuclear force. In less than half an hour, these divinely commissioned blueprints created enough hydrogen to provide the fuel for nuclear fusion in every future star!

This chapter is dedicated to the equation/blueprint, $E = mc^2$. Let there be light, let light create matter in mass-encoded quarks and protons, and launch a journey to galaxies, stars, Earth, and its life.

We Have Met the First Blueprint of Creation

$E = mc^2$

Car Fact

As noted in the introduction, we may initially view the universe as too complex to comprehend. But our car-savvy culture manages to manufacture complex automobiles en masse. The following is our initial "car fact":

Rather than using gasoline refined from fossil fuels, you could power your car with man-made liquid hydrogen (H2), close cousins of the protons that fuel the shining of the stars. Unfortunately, liquid hydrogen is amazingly inflammable and must be transported in special containers at ultra-low temperatures. NASA recently launched the first Artemis rocket in preparation for sending men and women back to the moon. The fuel for that rocket? Hydrogen: the same atom that fuels nuclear fusion in the stars.

Supplement One

In 1915, most scientists, Einstein included, thought the universe was eternal and stable. To describe an unchanging universe, Einstein added a "cosmological constant" to the equations of relativity. His universe would neither expand nor contract.

The astronomer-priest Georges Lemaître realized that Einstein's cosmological constant was a mistake. By deleting the constant, he recognized that the universe may be constantly expanding. Reversing the expansion, he reached a startling conclusion: the universe had a beginning, erupting from a "cosmic egg!"

Einstein resisted the notion, but Hubble's telescope confirmed the hypothesis. The universe began in a point of near-infinite heat and light, what we know today as the "big bang."

Question:
If the universe sprung from a "cosmic egg," who was the "cosmic chicken?"

CHAPTER TWO

Let There Be Mass

Photons of light have no mass. As a result, they set the speed limit of the universe. Photons can circle Earth seven times per second.

Like photons of light, quarks emerged from the light with no mass. Mass gives matter its heft, allowing material objects to occupy space and resist acceleration.[47] When the gravitational field of Earth acts on your mass, it determines your weight. A lawnmower might help explain mass. When you push a lawnmower up a hill, the lawnmower's mass resists the push. What if some miracle stripped away the lawnmower's mass? It would disappear into space at the speed of light.

If quarks remained mass free, they would continue to travel at light speed, ignoring their fellow quarks. But protons require teams of three quarks. If there were no mass, there would be no protons. If there were no protons, there would be no deuterium or tritium and no lights in the sky. Something had to give quarks mass, slow them down, and encourage them to work as a team.

The solution was embedded in the big bang in the form of a Higgs field, a mysterious field of energy filling the entire universe like no other field known to science. Higgs bosons, the particle of the

---

47   Lev B. Okun, "The Concept of Mass," *Physics Today* 42, no. 6 (June 1989): 31–36, https://doi.org/10.1063/1.881171.

Higgs field, fill the field like an ocean of invisible molasses. Quarks encounter that ocean, slow down, and acquire a perfect quantity of mass (see the definition of mass below).

Make the Universe Physical

The big bang delivered quantities of energy that defy our imagination. But God intended to fill creation with more than light. He intended to create an abundance of objects made of "matter." Galaxies and stars represent matter. So do planets and people.

In chapter 1, we met the vital equation $E = mc^2$ and recognized that matter (the m in the equation), is frozen energy. We also realized that protons were the most abundant form of newborn matter, storing energy for the stars and beyond.

The "standard model" of particle physics had a fundamental flaw. It failed to explain how electrons and quarks obtain mass. Defined as resistance to acceleration, mass gives material objects their heft. Gravity acts on our mass and translates that mass into weight. Without mass, quarks would fly into the expanding universe at the speed of light, producing nothing more than a light show.

How Would Electrons and Quarks, Components of the Atom, Acquire the Necessary Mass?

A magnificent blueprint rose to the challenge. A one-of-a-kind field, the Higgs field, producing particles known as Higgs bosons, instantly filled the newborn universe. Higgs bosons interacted with electrons and quarks, slowing them down and encoding them with just-right quantities of mass.[48]

How did scientists discover this mass-encoding mechanism? In one of the most dramatic developments in the history of science, physicists discovered the Higgs solution by analyzing a mathematical equation.[49] Should a field described by that equation fill the universe, they realized, it would accomplish the encoding of mass.

The Standard Model had a significant problem: There was no mechanism for mass that would make the universe physical.

Peter Higgs examined an equation and proposed a solution: If a field described by the equation filled the entire universe, it would provide the missing mass. High-energy particle collisions at CERN in Geneva Switzerland in 2012 confirmed the existence of a Higgs boson.

---

[48]   Juan Maldacena, "The Symmetry and Simplicity of the Laws of Physics and the Higgs Boson," European Journal of Physics 37, no.1 (October 2014): Article 015802, https://doi.org/10.1088/0143-0807/37/1/015802.

[49]   Peter W. Higgs, "Broken Symmetries and the Masses of Gauge Bosons," *Physics Review Letters* 13, no. 16 (October 1964): 508, https://doi.org/10.1103/PhysRevLett.13.508; Frank Close, *Elusive: How Peter Higgs Solved the Mystery of Mass* (New York: Basic Books, 2022).

In 1964, scientists proposed that a Higgs mechanism explains the origin of mass. In 2012, the Large Hadron Collider in Geneva, Switzerland, confirmed the existence of the Higgs boson.[50] Peter Higgs was right: the Higgs mechanism allows the universe to contain physical objects, including galaxies and stars and planets.

> **Question:**
> *Is the Higgs field any less a miracle than the manna of Exodus?*

The Higgs Mechanism

Every particle congealing from big bang light was traveling at the speed of light. Like the photons of light, they had no mass. Mass represents the "stuff" in an object that makes it resist acceleration. Mass is also the property of matter that gives physical objects their weight.

---

50   Celeste Biever, "It's a Boson! But We Need to Know if It's the Higgs," *New Scientist*, July 6, 2012, https://www.newscientist.com/article/dn22029-its-a-boson-but-we-need-to-know-if-its-the-higgs/.

Key Definition: Matter

A physical substance, as distinct from mind and spirit. Objects with matter occupy space and possess rest mass.

Key Definition: Mass

The property of resistance to acceleration. Think of a car sitting on the side of the road in neutral gear. You may have difficulty pushing that car a few feet. Its mass resists your push. The gravitational field of Earth acts on our mass to determine our weight. Want to weigh less? Go to the moon. In the smaller gravitational field of the moon, your weight will fall by 80 percent!

Key Definition: The Higgs field

A one-of-a-kind field filling the entire universe, producing particles known as Higgs bosons. Higgs bosons interact with electrons and quarks, encoding them with a just-right quantity of mass.

What's So Important About Mass?

Photons of light have no mass—and, therefore, they have no resistance to acceleration. As a result, they set the speed limit of creation, circling the Earth seven times per second.[51] Only particles with mass could slow down, interact with each other, and make something physical.

So, how did God assure that His universe would contain physical structure? He dispatched a powerful field (like all other blueprints, defined by a mathematical foundation) that could encode electrons and quarks with mass. Like no other field known to science, the

---

[51]   Wikipedia, s.v. "Speed of Light," last modified July 24, 2024, 00:36, https://en.wikipedia.org/wiki/Speed_of_light.

Higgs field and its bosons are distributed equally throughout the universe.[52]

Without this Higgs mechanism and the gift of mass, quarks would be zooming across an empty universe with no inclination to form protons or neutrons. The universe would have no physical structure, atoms, planets, or people.

A manger scene provides a metaphor for the loss. There would be no manger. There would be no sheep. There would be no Lamb of God and no stars to guide the wise men. And there would be no artist to paint the scene. The universe at large would be empty.

The Higgs field is a cosmological miracle, comparable to the "bread of heaven," "angel's food," or "spiritual meat." In the era of computers, we can recognize the Higgs field as something even more familiar: the Higgs is the motherboard of creation, encoding God's particles with mass.

If the Higgs Had Formed Differently

How important is the Higgs mechanism? Lawrence Krauss, former foundation professor at Arizona State University's School of Earth and Space Exploration and ex-director of the university's Origins Project, explains it well. "Every facet that is responsible for our existence, indeed the very existence of the massive particles from which we are made, would thus arise…the formation of…Higgs condensate [field].… The particular features that make our world what it is—the galaxies, stars, planets, people, and the interactions among all of these—would be quite different if the condensate [Higgs] had never formed. Or if it had formed differently."[53]

---

[52] Maldacena, "Symmetry and Simplicity."
[53] Lawrence M. Krauss, *The Greatest Story Ever Told—So Far* (New York: Atria Books, 2017), 217.

Putting the Higgs Mechanism in Perspective

We should ask this question at the end of each chapter: how many times do we uncover ingenious blueprints and dismiss them as accidents of nature?

Was the eruption of light in the primordial atom an accident? Did an accident give birth to quarks in the big bang light?

If these events were random accidents of nature, then it only makes sense that the Higgs field was accidental as well... right? Wrong.

We can rest assured that nature couldn't care less about a mass problem. Mother Nature would be perfectly content with a massless particle light show and a structure-free, non-physical world. It was a transcendent mind, not nature, who wrote the Higgs equation and endowed creation with mass.

We Have Met the Second Blueprint of Creation

$E = mc^2$
The Higgs field

Car Fact

Imagine attempting to drive a massless holographic version of your car. Your hand would effortlessly pass through the door handle. The steering wheel would be a wispy vapor. Higgs bosons encode the components of your car with mass. Slam the door, kick the tires, and be thankful for God's gift of mass.

Supplement Two

Learn more about the Higgs mechanism that gives the universe physical structure. In technical terms, the Higgs field makes the entire universe, in the words of Frank Wilczek, a "super-duper-super conductor," forcing electrons and quarks to navigate an ocean of Higgs bosons.[54]

---

[54] Frank Wilczek, "A Particle That May Fill 'Empty' Space," *Wall Street Journal*, December 29, 2022, https://www.wsj.com/articles/a-particle-that-may-fill-empty-space-11672337133.

CHAPTER THREE

Quark Code and a Strong Force

In chapter 1, we met the vital equation $E = mc^2$ and recognized that matter is "frozen" energy. We also learned that protons were the most abundant and useful form of newborn matter. Protons (hydrogen gas) deliver fuel to the stars.

In chapter 2, we reviewed a critical step in the production of protons. Massless quarks could hardly form protons as they zoomed into the universe at light speed. But a Higgs mechanism encoded electrons and quarks with a just-right quantity of mass. If the Higgs field was team player number one in the plan to convert quarks into protons, two new blueprints quickly joined the team.

A quark code organizes quarks into three-quark teams with just-right quantities of electric charge. Then, a force particle, the gluon of the strong nuclear force, enveloped those three-quark teams in its embrace, creating stable protons and neutrons for the nucleus of the atom. ‹

Let There be Electric Charge

We use electrons to light up our homes and charge our computers and smartphones. It's common knowledge today that electrons carry a negative, minus-one charge.

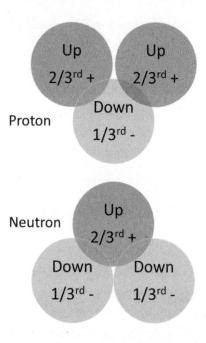

To their surprise, particle physicists in the 1960's discovered that protons and neutrons contain tiny particles inside with the strange name quark. Quarks come in two flavors, "Up" and "Down," and those quarks carry different fractions of electrical charge. As depicted in these images, quarks gather in three-quark teams: two "Up" quarks with one "Down" create a proton with a positive one charge matching the negative one charge of the electron. With just One "Up" quark and two "Down," neutrons are charge neutral.

Quarks also emerged from the light with charge, but their quantity of charge was strange: quarks possess fractions of electric charge. One flavor of quark, the up quark, has a two-thirds positive charge. Its cousin, a down quark, has a one-third negative charge.[55] Why create quarks with fractional charge? And why deploy them in teams of three? Chapter 3 will answer those questions and provide a striking example of divine design.

Code of the Quarks

We should analyze the math of three-quark teams. One team (two ups, one down) create protons with a net charge as follows: $2/3 + 2/3 - 1/3 = 3/3$. Protons have a positive electric charge. The plus-one (+1) charge of protons perfectly balances the negative-one (-1) charge of the electron.

A second three-quark team (one up, two downs) produces charge neutral neutrons: $2/3 - 1/3 - 1/3 = 0$. Charge neutral neutrons play a critical role during nuclear fusion in the stars by adding mass to a higher element without disturbing net charge.

The quark code would be meaningless without an additional blueprint called "the strong nuclear force," one of the fundamental forces of physics, locking quarks into protons and neutrons and assuring the stability of atoms.[56]

---

[55] M. Y. Han, *Quarks and Gluons: A Century of Particle Charges* (Singapore: World Scientific Publishing, 1999).

[56] Josh Bastianello, *The ABC Pocket Guide to the Standard Model of Particle Physics* (self-pub., Amazon Digital Services, 2020), 9.

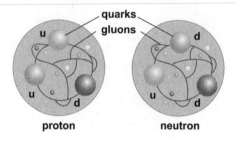

quarks
gluons
u
d
u
d
u
d

proton   neutron

Gluons of the Strong Nuclear Force lock quarks in place
inside protons and neutrons in the nucleus of the atom.
Gluons act like furiously vibrating rubber bands, assuring
that quarks never exist freely on their own.

Question:

*Do these fractional quark charges seem arbitrary and
accidental?*

What Creates Matter?

Atoms make matter, but that answer is incomplete. A better response:
protons and neutrons form the nucleus of atoms, while electrons
orbit overhead. That answer is correct but also incomplete. What
really creates matter? Particles called quarks join in teams of three
to form the protons and neutrons of the atom. *Quarks lie at the
heart of matter.*

Why do quarks work in teams of three? Their teamwork
exploits their fractional charge, giving protons a positive charge of
plus one, precisely balancing the negative charge of the electron.

Enter the Strong Nuclear Force

Quarks are slippery creatures. Something had to glue quarks into three-quark teams to stabilize the proton and nucleus of future atoms. The strong nuclear force provides the stabilizing factor. Its force particle, the aptly named gluon, locks them securely in place.[57] Thanks to the strong nuclear force, quarks never exist freely in nature but are always locked tightly inside the protons and neutrons of the atom.[58]

The Quark Code is a Striking Example of Superintelligent Design

Francis Collins, the former director of the National Institutes of Health, expresses his impression of quark code in his well-received book, *The Language of God*, stating, "For those who argue that materialism should be favored over theism, because materialism is simpler and more intuitive, these new concepts present a major challenge."

He continues: "Their mathematical representation invariably turns out to be elegant, unexpectedly simple…. Why should matter behave in such a way?"[59]

The widely respected (and unfortunately recently deceased) science writer, Amir Aczel, recorded his impression of the quark code in his 2014 book, *Why Science Does Not Disprove God*: "I have marveled myriad times about what to me is one of the greatest

---

[57] Wouter Schmitz, *Particles, Fields and Forces: A Conceptual Guide to Quantum Field Theory and the Standard Model* (Cham, Switzerland: Springer, 2019), 133.

[58] Michael Riordan, *Hunting of the Quark: A True Story of Modern Physics* (New York: Simon and Schuster, 1987).

[59] Francis S. Collins, *The Language of God: A Scientist Presents Evidence for Belief* (Rockland, MA: Wheeler Publishing, Inc., 2007), 61, 62.

mysteries of all: how, within the immensely hot and dense 'soup of particles' that constituted our universe, a fraction of a second after the Big Bang, the quarks suddenly gathered in threes: two 'ups' and a 'down' to form protons and two 'downs' and an 'up' to form neutrons. How was it ever possible, I have asked myself, that the charges of these quarks turned out to be exactly 2/3 for an 'up' and -1/3 for a 'down'—with not even the tiniest of errors, so that the proton would miraculously match the opposite charge of the electron (-1) and the neutron's charge would be precisely zero: just what is necessary to form atoms and molecules? How did such an incredibly improbable event ever happen without some calculated act of creation? And further, how did the masses of the elementary particles turn out to have the perfectly precise ratios needed so that our world of atoms and molecules could exist at all?"[60]

These two leading observers recognize the implications of quark code:

A superintelligent mind endowed quarks with fractional electric charge and dispersed them in three-quark teams.

We Have Met the Third and Fourth Blueprints of Creation

$E = mc^2$

The Higgs field

Quark code

Gluons and the strong nuclear force

---

[60]   Amir D. Aczel, *Why Science Does Not Disprove God* (New York: Harper Collins, 2014), 2.

Car Fact

If your car functioned like the universe, its carburetor would use the quark code and gluons of the strong force to convert nonflammable massless quarks into energy-rich hydrogen protons.

After three successive miraculous interactions with

1. the Higgs field for mass,
2. other quarks to create three-quark teams, and
3. the strong force to stabilize the process, the clouds of hydrogen fuel were ready for a journey to the stars.

Supplement Three

Quarks create protons and neutrons for the nucleus of the atom. Beginning with Democritus in ancient Greece, human beings have struggled to understand the components of the atom.

CHAPTER FOUR

Quantum Creativity

The creation of the universe involved untold numbers of mergers and acquisitions as fundamental particles created atoms, heavier elements, and the chemical basis for life.

At every step along the way, particles displayed magical superpowers, exploring multiple locations in the same instant (superposition) and forming lasting relationships with other particles across great distances of space (entanglement).

Creative Features of the Quantum World

In earlier chapters, we reviewed the ingenious energy management program in the newborn universe. The big bang stored vast quantities of energy in the protons of hydrogen gas. We recognized that protons are the composite product of strange particles called quarks. Remember, a quark code determines that two up quarks and one down quark form a proton, while one up and two downs create a neutron. We will soon see how protons fuel nuclear fusion in the stars.

Superposition

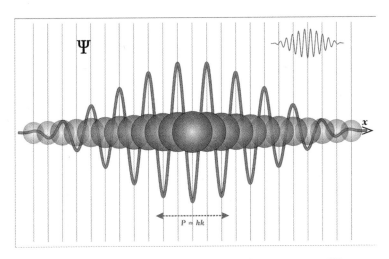

As illustrated by this image, a single particle can occupy trillions of locations in the same instant of time. Imagine the impact of this superposition during nuclear fusion in the stars and the construction of many features of our vast and beautiful universe.

This chapter celebrates the quantum nature of particles that makes nuclear fusion possible. If the big bang produced particles that resembled tiny marbles, bouncing off each other like billiard balls, the universe would have been stillborn. God maximized their creativity by allowing particles to behave like probability equations, essentially cloning themselves to explore the universe in a quantum state called a superposition.[61] Until the superposition is disturbed by a measurement (such as a scientific observation or collision with

---

61   Jim Al-Khalili, *Quantum: A Guide for the Perplexed* (London: Weidenfeld and Nicolson, 2012), 24–25.

a potential destination), quantum particles explore multiple paths before committing to a time and place.

Entanglement

God also empowered particles to form coalitions through a process called entanglement. Once entangled, particles can face a shared future.[62] When one particle entangles with another, those two particles function as a seamless system, making a commitment that defies location and time. Entangled particles maintain contact, even when separated by thousands of miles.[63] Einstein derided this concept as "spooky action at a distance,"[64] but experiments such as those conducted by space satellites described below have confirmed that quantum entanglement is a fundamental feature of creation.[65]

The Quantum Wave Function

$$i\hbar \frac{\partial}{\partial t} \Psi = \hat{H} \Psi$$

The quantum wave function describes the probability that a particle will occupy a given location and possess a certain quantity of spin.

---

[62] Amir D. Aczel, *Entanglement: The Greatest Mystery in Physics* (New York: Four Walls, Eight Windows, 2002).

[63] Brian Clegg, *The God Effect: Quantum Entanglement, Science's Strangest Phenomenon* (New York: St. Martin's Press, 2009).

[64] Walter Isaacson, *Einstein: His Life and Universe* (New York: Simon and Schuster, 2007), 448.

[65] Alain Aspect, Philippe Grangier, and Gérard Roger, "Experimental Realization of Einstein-Podolsky-Rosen-Bohm Gedankenexperiment: A New Violation of Bell's Inequalities," *Physical Review Letters* 49, no. 2 (July 1982): 91–94, https://doi.org/10.1103/PhysRevLett.49.91.

An equation called the quantum wave function describes the location, spin, mass, and other attributes of quantum particles, broadcasting the availability of each particle to engage in new relationships.[66]

Quantum Particles Also Display Two Features

They have a dual nature—they travel as waves (best understood as probability equations) but gather themselves into discreet particles at their destination.

They cloak their properties in uncertainty: extract any information from a particle (such as its mass, location, or spin), and you give up the ability to recover further information about the particle.

Question:
Could energy become protons, atoms, stars, and a table of elements without the powers of superposition and entanglement?

The Development of Quantum Mechanics

The enormous difference in scale between our human experience and a particle smaller than an atom challenges our imagination. It is hard to understand anything that microscopic. Even more remarkable, the smallest units of creation play by quantum rules. The classical world confines us to a specific location at a well-defined moment in time. Not so for particles in the quantum domain.[67]

---

[66] Daniel A. Fleisch, *A Student's Guide to the Schrödinger Equation* (Cambridge: Cambridge University Press, 2020).
[67] Al-Khalili, *Quantum: A Guide.*

From photons and electrons to atoms and molecules, the building blocks of creation have the potential to explore multiple locations in the same instant. They can entangle with fellow particles and maintain that relationship across deep space. All the physical features of our planet and its life-forms reflect the cumulative aggregation of these tiny vibrations of energy, which defy classical concepts of distance and time.

Max Planck was a leading German physicist during WWII, discovering the fundamentals of quantum mechanics and admonishing medical students to combine their medical skills with religious faith.

Max Planck was Germany's leading scientist during WWII. During the war, he delivered lectures to medical students across Germany, imploring them to embrace faith as well as science, to view them as complementary views of reality.[68]

As a physicist, Planck studied the theoretical release of heat energy from an idealized black box or stove.[69] Common sense

---

[68] Brandon R. Brown, *Planck: Driven by Vision, Broken by War* (Oxford: Oxford University Press, 2015).
[69] Brown, *Planck: Driven by Vision*.

suggested that the heat would escape slowly and smoothly as the black box gradually cooled. Instead, the energy flowed like a marble bouncing down steps, reaching cooler plateaus in staccato fashion, stepping down to one lower level after another. Planck discussed these unexpected results with a rising star in German physics.

Who did you call in 1905 when you confronted a surprise in physics? Max Planck called Albert Einstein. Together, they analyzed heat equations and launched a new field called "quantum mechanics."[70] Heat energy did not flow from the black box as a waterfall. It bounced. It was quantized or packaged. The heat took "quantum jumps," determined by the wavelength of its starting energy.

In their native state, particles are anything but hard little BBs. Instead, they are minuscule ripples of energy, spread out in traveling wavefronts described by an equation called the "quantum wave function."[71] The wave function provides a snapshot of the particle's potential behavior: will it spin up or down, have this much mass, or have that much electric charge? What are its possible physical locations? Think of the wave function as an "all-points bulletin" or "most-wanted poster" outlining the particle's potential.

You may be familiar with the term "quantum leap," popularized in modern culture. Packets of vibrating energy can indeed leap to higher resonance levels, much like you can play piano notes in higher octaves. But they cannot take random leaps. An atom's electron can absorb energy and jump to the next highest orbit, but a wave function defines that orbit.

---

70 A. Douglas Stone, *Einstein and the Quantum: The Quest of the Valiant Swabian* (Princeton: Princeton University Press, 2013), 3.
71 Loew T. Kaufmann, *Quantum Physics for Beginners* (self-pub, Amazon Digital Services, 2021), 34.

A Particle of Creative Potential

The wave function is a description of the particle's creative potential. It is a commandment to go forth and enter relationships. If the particle is a quark, its job is to team up with other quarks and create a proton or a neutron. If the particle is a proton or neutron, its mission is to join with other protons and neutrons and build the nucleus of an atom. If the particle is an electron, it may find a lonely nucleus and make a complete atom. If the particle is an atom, it may join with other atoms to produce molecular combinations.

The Most Important Experiment in Physics

The classic double-slit experiment demonstrates the ability of particles to exist as both particles and waves.[72] Shine a beam of light (or electrons or molecules) through a barrier with two slits and observe the pattern that develops on a photographic screen. Rather than creating two bright lines—as you would expect if particles flowed like bullets through the individual slits—the particles produce a pattern of alternating dark and light bands.

---

[72] John Gribbin, *Six Impossible Things: The Mystery of the Quantum World* (Cambridge, MA: MIT Press, 2019), 8.

Diffraction Light
Double-slit experiment (Young wave theory)

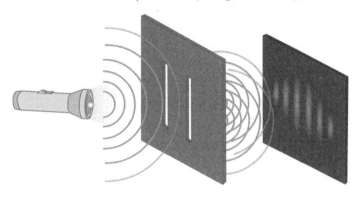

One of the best demonstrations of the dual nature of light, the
"Dual Slit" experiment demonstrates that light travels as a wave but
arrives at its destination as a particle. Note the interference pattern
on the screen as waves rise together or cancel each other out.
Bands on the screen are created by the particle behavior of light.

These particles travel as waves. Sometimes the waves rise
together, creating brighter bands. At other points, they cancel each
other out, producing dim bands. This wave interference pattern
reveals a fundamental truth: the particles first travel as overlapping
waves before gathering themselves as particles when they reach
their destination (a photographic plate, in the typical experiment).

What happens if we send photons, electrons, or other quantum
particles through the slits one particle at a time? An interference
pattern still appears on the screen.[73] Individual particles interfere
with themselves. They have used their wave function like a Ful-
ly-funded, unrestricted airline ticket to travel through both slits

---

[73] Al-Khalili, *Quantum: A Guide*, 194.

before committing to a final location. Each particle explores the full range of options enshrined in its wave function, exploring every path as they travel into the future.

Scientists have completed double-slit experiments with molecules of increasing size. One team in Vienna performed the dual-slit experiment with "buckyballs," a molecule containing sixty carbon atoms resembling a geodesic dome.[74] Despite containing four hundred protons, four hundred neutrons, and four hundred electrons, individual buckyballs passing through two slits can still clone themselves into multiple copies, spread out into numerous wavefronts, and produce an interference pattern.

As summarized by the Vienna team: "All these observations support the view that each C60 [carbon-60] molecule interferes with itself only."[75]

Amazing Fact:
Even when dispatched one molecule at a time, buckyballs travel through both slits and create an interference pattern. A sixty-carbon molecule enters a superposition, explores each available path and interferes with itself.

More recently, physicists have demonstrated quantum interference with molecules of gramicidin, a natural antibiotic made up of fifteen amino acids.[76] Their work raises a fascinating question: do enzymes, DNA, and even simple biological forms such as viruses,

---

[74] Markus Arndt et al., "Wave-Particle Duality of C60 Molecules," *Nature* 404, (1999): 680–682, https://doi.org/10.1038/44348.
[75] Markus Arndt et al., "Wave-Particle Duality."
[76] A. Shayeghi et al., "Matter-Wave Interference of a Native Polypeptide," *Nature Communications* 11, (2020): article 1447, https://doi.org/10.1038/s41467-020-15280-2.

exploit the power of quantum mechanics? Do quantum effects drive DNA mutations or the flow of thoughts in our brain?

Entangled in Space

Increasing quantum creativity even further, two wave functions can entangle and journey into the future as a single system.[77]

China launched the world's first quantum satellite, Micius, in 2016.[78] Orbiting one thousand miles above Earth, the satellite directed beams of entangled photons at ground-based telescopes separated by more than seven hundred miles. Both particles began their journey in a superposition. When one ground station measured one of the entangled photons, the second photon, now hundreds of miles away, instantly abandoned its superposition.[79] The entangled partner also displayed properties "complementary" to the measured photon. Upon measurement, if the particle was spinning left, for instance, its entangled partner would immediately spin right.

Two particles had traveled into the future as one entangled system, defying hundreds of miles of physical separation. It seems unreal, but quantum theory has been invaluable in helping us understand the world.

---

[77] Amir D. Aczel, *Entanglement: The Greatest Mystery.*

[78] Karen Kwon, "China Reaches New Milestone in Space-Based Quantum Communications," *Scientific American*, June 25, 2020, https://www.scientificamerican.com/article/china-reaches-new-milestone-in-space-based-quantum-communications/.

[79] Gabriel Popkin, "China's Quantum Satellite Achieves 'Spooky Action' at Record Distance," *Science*, June 5, 2017, www.science.org/content/article/china-s- quantum-satellite-achieves-spooky-action-record-distance.

New Perspectives on Reality

In conjunction with the theory of relativity, the development of quantum mechanics destroyed the following presumptions about nature:

Particles and waves are fundamentally different.

No, any particle can also behave as a wave.

Time is the same for all observers.

No, light is constant. Time and distance are relative.

Nature is predictable.

No, nature rides on quantum rails and is fundamentally unpredictable.

With sufficient knowledge, we can forecast how systems will evolve with certainty.

No, the best we can do is predict probabilities.

A Skeptical Einstein

Although Albert Einstein helped Max Planck launch quantum mechanics, describing that heat energy flows in discreet amounts called "quanta," Einstein never warmed up to quantum concepts. In a 1926 letter to a quantum physicist, he delivered a famous opinion: "Quantum mechanics is very impressive. But an inner voice tells me that it is not yet the real thing. The theory produces a good deal but hardly brings us closer to the secret of the Old One. I am at all events convinced that *He* does not play dice."[80]

Einstein's Skepticism Aside, Quantum Mechanics Changed the Face of Physics

For those who find quantum mechanics challenging to understand, one of the great physicists, Richard Feynman, offers some solace: "I think I can safely say that nobody understands quantum mechanics."[81]

The quantum features of the microscopic world reveal fundamental truths: God endows creation with free will, and the quantum nature of reality represents one of His most powerful tools.

We Have Met the Fifth Blueprint of Creation

$E = mc^2$

The Higgs field

Quark code

Gluons and the strong nuclear force

Quantum superposition and entanglement

---

[80] Walter Isaacson, *Einstein: His Life and Universe*, (Simon and Shuster Paperbacks, 2007) 334.

[81] Richard Feynman, "Probability and Uncertainty: The Quantum Mechanical View of Nature," 1964, in Messenger Lectures at Cornell University, California Institute of Technology and The Feynman Lectures Website, transcript and HTML5 video, at 8:10, 56:32, https://www.feynmanlectures.caltech.edu/fml.html#6.

Car Fact

Imagine your car could express the quantum nature of its subatomic components: existing in a superposition, it would fill your garage and driveway with identical cars. Then it would entangle with your neighbor's SUV, displaying complimentary odometer readings no matter the distance between them.

Supplement Four

Life is a product of carbon. Carbon creates DNA, proteins, and other biomolecules that constitute 18 percent of our body. To assure an adequate supply of this vital element, God endowed carbon with a perfect quantum leap to double its production in the stars.

CHAPTER FIVE

Space-Time

Space-time provided a container for the energy, particles, and forces flowing from the big bang. It also provides the gravitational squeeze necessary to light up the stars. The warping of space-time makes objects with mass attract each other (actually, fall toward each other down the slopes of space-time), as space-time constantly recalibrates the large-scale structure of the universe.

Multiple Roles for Space-Time

Energy sufficient to produce more than five-billion-trillion stars erupted from a point smaller than an atom—a singularity, in the language of modern physics.[82] Space-time was a crucial partner in the process, providing a container for the big bang eruption.

As the Higgs mechanism encoded quarks with mass, and gluons snapped teams of quarks into protons, enormous clouds of hydrogen gas floated onto the space-time fabric. Encoded with mass, those hydrogen clouds (vast collections of protons) disturbed the delicate geometry of space-time. In response to this space-time warping, gravity surrounded those clouds in its squeeze, shaping them into round, hot spheres. As the temperature and compression

---

[82]   Roger Penrose, *The Road to Reality: A Complete Guide to the Laws of the Universe* (New York: Vintage Books, 2004), 130.

reached a threshold, stellar light defeated the darkness. God said, "Let there be light." Then He created trillions of lights in the sky.

Cosmic Scaffolding

Space-time became cosmic scaffolding as dramatic events took place on its stage. Ancient stars died in massive supernova explosions,[83] while other stars collapsed into black holes.[84] Galaxies collided and merged, creating gargantuan galactic clusters.[85]

Would these cataclysmic events make the geometry of the heavens fly apart, creating chaos across the cosmos? To the contrary, the gravitational fabric took it all in stride, constantly readjusting the distribution of mass. New stars are born, older stars die, and galaxies merge with one another. The space-time vessel plays juggler-in-chief, which assured smooth travels for a special planet that would one day become home for living creatures.

Question:
Scientists remain divided over the nature of space-time: is it a fundamental feature of our universe or evidence that we live in a hologram?

---

[83] Anna Frebel, *Searching for the Oldest Stars: Ancient Relics from the Early Universe* (Princeton: Princeton University Press, 2015), 51.

[84] Brian Clegg, *Gravitational Waves: How Einstein's Spacetime Ripples Reveal the Secrets of the Universe* (London: Icon Books, 2018), 32.

[85] Yvette Cendes, "How do Black Holes Swallow Stars?" *Astronomy*, updated December 29, 2021, https://www.astronomy.com/science/how-do-black-holes-swallow-stars/; Matthew A. Malkan and Ben Zuckerman, eds., *Origin and Evolution of the Universe: From Big Bang to Exobiology*, 2nd ed. (Singapore: World Scientific, 2020), 19.

Space–Time: A Multifaceted Blueprint

Different reviewers have described space-time as a container, vessel, scaffolding, fabric, and stage for cosmic events. What do all these metaphors imply? Space-time has structure. Space-time plays an active role in the cosmic drama.

It required more than one hundred winners of a Nobel Prize to construct the standard model of particle physics. It took one human being to explain gravity. Albert Einstein mentally rode a beam of light and realized that four-dimensional space-time fills all of space like a multilayered spider web. Objects with mass warp and indent the spider web, causing every object with mass to fall into the warp. The result: objects with mass attract (fall toward) each other.[86]

As Einstein surmised in a thought experiment, gravity reflects the warping of the space-time fabric by objects with energy or mass. Here Earth warps its surrounding space-time and determines the orbit of the moon.

---

[86] Albert Einstein, *Relativity: The Special and the General Theory - 100th Anniversary Edition* (Princeton: Princeton University Press, 2015).

Before Einstein, humankind gave little thought to space as an active component of creation. Space seemed like an empty backdrop with no structure, just a passive canvas for the stars. Ancient astronomers studied constellations, and wise men monitored comets for messages from the gods.

The next time you walk down a staircase, notice how easily you could fall. Then, feel the energy required to climb back up the stairs. In everyday life, we constantly feel the power of space-time and the distortion of space-time by Earth. And yet, we have no clear understanding of the substance and structure of space-time. Wave your hand through the air. Did you feel the nodes of space-time? How about the bosons of the Higgs field? They occupy our universe in unimaginable numbers, surrounding us with divine mystery.

Isaac Newton

In the 1700s, an apple dropped in an orchard, motivating Isaac Newton to recognize gravity as a force of attraction between objects with mass. The gravitational power of Earth, he realized, controlled the fall of the apple.

Notably, Newton was the first scientist to support his theory with mathematical equations. Every object embedded in space-time attracts every other object in proportion to their mass. This gravitational attraction decreases as objects grow further apart. When space-time is flat, Newton's equations are valid today.[87] But when a star or planet or black hole warps the geometry of space-time—when strong gravitational fields convert flat space-time into deep indentions—Newton's equations lose their explanatory power. Einstein recognized Newton's misconceptions. Space-time

---

[87]   James Gleick, *Isaac Newton* (New York: Vintage Books, 2003).

was no passive backdrop. It plays an active role in the cosmic drama as objects with energy or matter curve the space-time continuum.

Theory of Relativity

Einstein's theory of relativity not only explained gravity as the warping of space-time but also changed our understanding of distance and time. While Newton saw distance and time as absolute features of the universe, Einstein made them "relative," dependent on the observer's perspective.[88] Light was the constant of creation, moving at unchanging speed. Whether standing still or zooming through space on a rocket, an observer would always find light speeding away at 186,000 miles per second. The speed of light is a fundamental constant of physics represented by the letter c (the c in $E = mc^2$).[89]

Think you want to catch a beam of light? Good luck. As you race after the light, you grow heavier. Closing in on the light, your mass increases. The hands on your watch begin to slow down. The distance you once thought was a foot soon measures an inch.

Nothing can catch the light. Objects with mass, be they spaceships or people, would need an infinite supply of energy to match the speed of light.[90]

---

[88] Richard Wolfson, *Simply Einstein: Relativity Demystified* (New York: W. W. Norton and Company, 2003), 18.
[89] Walter Isaacson, *Einstein: His Life and Universe* (New York: Simon and Schuster, 2007), 139.
[90] Alok Jha, "Why You Can't Travel at the Speed of Light," *The Guardian*, January 12, 2014, https://www.theguardian.com/science/2014/jan/12/einstein-theory-of-relativity-speed-of-light.

Team Travel

**The gravitational field of the sun determines
the orbits of its eight trailing planets.**

Once objects with mass develop relationships based on their gravitational attraction, they travel through space as a team.[91] Our sun belongs to the gravitational field of the Milky Way, slowly orbiting its galactic center. Our Earth is a member of team solar, one of eight planets orbiting the sun.

More to the Story: Putting
Space-time to the Test

Einstein designed the first experiment to test his theory of relativity. He predicted that the sun's gravitational field would bend the path of light traveling from a distant star. The light would never slow,

---

[91]   Jim Bell, *Hubble Legacy: 30 Years of Discoveries and Images* (New York: Sterling 2020).

but the strong gravitational field of the sun would distort the space-time surrounding the sun, redirecting the starlight's path. If you could record the star's position during the undisturbed nighttime sky and record it again during a daytime eclipse, observers would see the star in two separate locations.

On March 29, 1919, British astronomer Sir Arthur Eddington traveled to Principe Island, off the western coast of Africa, where his team photographed stars during a solar eclipse and again in the nighttime sky.[92] Just as Einstein predicted, the apparent location of the star had shifted. The gravitational field of the sun had warped the geometry of space-time and curved the trajectory of light.

Overnight, Einstein and the theory of relativity became a worldwide sensation.

We Have Met the Sixth Blueprint of Creation

$E = mc^2$
The Higgs field
Quark code
Gluons and the strong nuclear force
Quantum superposition and entanglement
Space-time

Car Fact

Too bad your car uses Earthbound highways rather than the autobahn of space-time. Earth travels through space at sixty-seven thousand miles per hour, while our galaxy, the Milky Way, speeds through the universe at more than one million miles per hour!

---

[92]   Isaacson, *Einstein: His Life and Universe*, 257.

Supplement Five

The geometry of the space-time continuum produces gravity and constantly recalibrates the large-scale structure of the universe. If space-time functions like a multilayered fish net, what knits the nodes of space-time together? Do the bits of a quantum computer or "qubits" tie the fabric of space-time together?

Scientists have proposed an even more startling hypothesis. Is our universe a hologram? Like the black holes at the heart of every galaxy, does our universe have an "event horizon," a lower-dimensional quantum field resembling a CD that records every object and event in the universe? In that case, space-time would not be fundamental but a derivative product entangled with that quantum field. As Leonard Susskind, professor of theoretical physics at Stanford University and founding director of the Stanford Institute for Theoretical Physics, first declared in the 1990s, "The three-dimensional world of ordinary experience—the universe filled with galaxies, stars, planets, houses, boulders, and people—is a hologram, an image of reality coded on a distant two-dimensional surface."[93]

---

[93] Leonard Susskind, *The Black Hole War: My Battle with Stephen Hawking to Make the World Safe for Quantum Mechanics* (New York: Little, Brown and Company, 2008), 410.

CHAPTER SIX

Nuclear Fusion

In the first twenty minutes of creation, the newborn universe functioned as a fusion reactor, storing vast quantities of quarks in the protons of hydrogen gas. But life would need more than hydrogen. It would need heavier elements and their molecular combinations. The blueprints of creation rose to the challenge, scattering nuclear reactors called stars across the heavens. Stars produce the raw materials for life.

Even though the energy of the big bang defies our imagination, its light has gradually faded. Today, scientists detect the faint remnant of that energy as the cosmic microwave background, first detected as background static on radios and television.[94]

In the absence of that primordial energy, how would God light up the Earth, providing energy for "the fish of the sea…the fowl of the air…the cattle…and every creeping thing that creepeth upon the earth"?[95] He deployed a powerful energy program known as nuclear fusion in the stars.[96] We have reviewed the creation of

---

[94] Erik M. Leitch, "What is the Cosmic Microwave Background Radiation?" *Scientific American*, November 1, 2004, https://www.scientificamerican.com/article/what-is-the-cosmic-microw/.

[95] Genesis 1: 20–24 (NIV).

[96] Arthur Turrell, *The Star Builders: Nuclear Fusion and the Race to Power the Planet* (New York: Scribner, 2021).

protons (mass-encoded quarks glued in teams of three). The time has arrived for protons to light up a star.

Protons and the Strong and Weak Force

At the start of the process, gluons of the strong nuclear force in the star hold two protons close together (both protons have a positive charge, and particles with the same charge naturally repel each other). With two protons forced shoulder to shoulder, W particles of the weak nuclear force plunge into one of the up quarks in one of the two protons, converting one of the two up quarks in that proton to a down. A proton containing two up quarks and one down instantly becomes a neutron with only a single up quark and two down. Two unrelated protons suddenly become a stable proton-neutron combination, a new element or "isotope" of hydrogen known as "deuterium."[97]

From Deuterium to Tritium

Then the process completes a second round. The strong force brings another proton into proximity with the proton of deuterium. The weak force repeats its action, creating yet another form or isotope of hydrogen containing a single proton with two neutrons, called "tritium."[98]

---

[97] Michio Kaku, *The God Equation: The Quest for a Theory of Everything* (New York: Doubleday, 2021), 92.
[98] Turrell, *The Star Builders*, 42.

Release a Fraction of the Starting Mass as Energy

The last step in stellar fusion is a spectacular event. Deuterium (one proton and one neutron) fuses with tritium (one proton and two neutrons) to form helium (two protons and two neutrons). The process ejects the excess neutron and returns 0.07 percent of the starting mass to its native state as pure energy.[99] Einstein's famous equation explains the power of the process: that small amount of matter (m) returning to energy (E) is multiplied by the speed of light squared (c^2), an enormous eighteen-digit number.

> **Question:**
> *The W particle of the weak nuclear force was a component of the newborn universe yet played its most important role much later during the creation of the stars. Does its inclusion in the "big bang" package seem like serendipity or a reflection of divine planning before God declared the first "Let there be..."?*

Humanity's First Involvement with the Power of the Atom

Human beings' first encounter with atomic energy occurred during WWII, as both sides rushed to develop a nuclear weapon. That weapon would exploit atomic fission by splitting a heavy atom and releasing its energy stores.[100] Uranium is a large, unstable atom. In

---

99   Turrell, *The Star Builders*, 6.
100   Cynthia C. Kelly, ed., *The Manhattan Project: The Birth of the Atomic Bomb in the Words of Its Creators, Eyewitnesses, and Historians* (New York: Black Dog and Leventhal Publishers, 2020).

an atom bomb, scientists bombard uranium (the matter or the *m* of the equation, in this instance) with neutrons to trigger a chain reaction. The equation runs from right to left, $mc^2 = E$, returning matter to energy. The result is a mushroom cloud with enough power to destroy a major city.[101]

Building a Star on Earth

Scientists are working hard to build a fusion reactor on Earth, and they claim to be making progress.[102] The number one challenge to their effort? They are struggling to produce an adequate supply of tritium. Tritium is a radioactive form of hydrogen that rapidly deteriorates on Earth. Compare their struggle to obtain sufficient tritium to the effortless production of tritium by nuclear forces in the stars.

Fusion Versus Fission

The goal of nuclear fusion is vastly different: recreating the nuclear power of the sun by forcing light elements to fuse. In nuclear fusion, gravity and enormous heat and pressure squeeze simple atoms together. Why would that trigger a release of energy? Because a small amount of the starting material, a small portion of deuterium and tritium, returns to its native form as pure energy. Helium, the product of nuclear fusion in our sun, weighs 0.07 percent less than the starting materials that combine to produce it.

---

[101] Kelly, *The Manhattan Project.*
[102] Philip Ball, "The Race to Fusion Energy," *Nature* 599 (November 2021): 362–366, https://www.nature.com/immersive/d41586-021-03401-w/index. html; Daniel Clery, "Laser-Powered Fusion Effort Nears "Ignition," *Science* 373, no. 6557 (August 2021): 841, https://www.science.org/doi/epdf/10.1126/ science.373.6557.841.

The sun "burns" about 564 million tons of hydrogen per second, producing 559.7 tons of helium. What happens to the lost matter? It is multiplied by a factor equal to the speed of light squared. Lest this sound like a pittance, remember the multiplying factor (the speed of light squared): 299,792,458 meters per second.[103]

As the star ages, additional fusion pathways convert helium into carbon, oxygen, and nitrogen. Later in its lifespan, the star may go supernova, exploding with enough intensity to complete the fusion of the heaviest elements.[104]

Time to Review and Reflect

Quarks have been at the center of creation on two separate occasions.

1. In the first moments of creation, quarks stored enormous big bang energy in almost unlimited clouds of hydrogen gas. As noted above, our sun consumes almost six hundred million tons of hydrogen per second. Remember, the production of hydrogen fuel for our sun happened in the first twenty minutes of creation.

2. Another fact to remember: to prepare quarks for their mission, God used the Higgs mechanism to encode them with mass and the strong force to lock them into protons and neutrons. What amount of time was required for this preparation? Less than twenty minutes.

3. Eons later, quarks deliver an amazing return on investment. The weak force converts a proton with two up quarks and one down into a neutron with only one up quark and two

---

[103] David Bodanis, *E=mc²: A Biography of the World's Most Famous Equation* (New York: Berkley Publishing Group, 2000), 70.
[104] Laurence A. Marshall, *The Supernova Story* (Princeton: Princeton University Press, 1988).

down. Freestanding protons become two forms of the hydrogen atom known as deuterium and tritium, the perfect fuel for nuclear fusion in the stars.

The twofold punch of the strong and weak nuclear force during nuclear fusion stands with $E = mc^2$, the Higgs equation, the quantum wave function, the quark code, and the theory of relativity as thoughts from the mind of God. If you think they are accidents of nature, be sure to ask the author about a lovely bridge for sale, connecting Tennessee with the fertile fields of the Arkansas delta. You can probably negotiate a discount.

We Have Met the Seventh Blueprint of Creation

$E = mc^2$
The Higgs field
Quark code
Gluons and the strong nuclear force
Quantum mechanics
Space-time
Nuclear fusion and the weak nuclear force

Car Fact

If your car's engine could function like a nuclear fusion reactor, you would enjoy the ultimate pollution-free clean energy. Unfortunately, the acquisition of tritium on Earth costs $30,000 per gram.

Supplement Six

Multiple labs around the world hope to create a fusion reactor on Earth. How will they replace the gravitational squeeze of the sun, which is four hundred times more massive than Earth? And as noted above, how will they produce sufficient tritium fuel?

CHAPTER SEVEN

Electrons and the Electromagnetic Force

In the newborn universe, blueprints of creation produced enough hydrogen to fuel trillions of future stars. But God wanted more than lights in the heavens. He wanted complex chemistry capable of animating life. How would He produce a table of elements and life-promoting molecules, including a just-right universal solvent (water, the only liquid that floats upon freezing) and organic biochemicals, including proteins and DNA? You will soon realize the Maker of the Heavens and the Earth is the Chemist-in-Chief of creation.

As scientists began to identify the characteristics of heavier elements in the eighteenth and nineteenth centuries, they started to recognize patterns. The result was a table of elements.[105] The table organizes elements by their atomic number—the number of protons in their nucleus. Hydrogen contains a single proton, and its atomic number is one. The heaviest elements in the table have more than one hundred protons.

---

[105] Ben Chinnock, *The Elements of the Periodic Table: The Complete Handbook* (Buckinghamshire, UK: B. C. Lester Books, 2020).

Each horizontal row on the periodic table is called a period. As you move from left to right across periods, elements that display metal properties become less and less metallic.

Each vertical column on the periodic table is called a group. Elements belonging to one of the eighteen groups will have similar properties. The halogens, including fluorine, chlorine, and bromine, provide a good example.

Most stars in our sky fuse deuterium and tritium to produce helium. But when fuel for this process is exhausted, stars more massive than the sun conduct fusion in an outer shell, forming oxygen, nitrogen, and heavier elements. Conventional fusion can produce elements as heavy as iron. But it requires the extreme astrophysical event of a supernova explosion during the death of a massive star to create the heaviest elements.[106]

Question:
As elements erupted from the stars, they were quickly joined by electrons orbiting in shells. Was it simply fortuitous that atoms could share electrons and form life-friendly molecules such as the water (H_2O) that makes up more than half of the human body?

Bring on the Electromagnetic Force

A moment of glory has arrived for the electron and the electromagnetic (EM) Force. The electron has been patiently waiting for stars

---

[106] Priyamvada Natarajan, *Mapping the Heavens: The Radical Scientific Ideas That Reveal the Cosmos* (New Haven: Yale University Press, 2017).

to disgorge their elements to complete the production of atoms.[107] Once electrons orbit the atomic nucleus, atoms can form larger molecules. How do atoms form molecules? They share electrons, creating numerous biofriendly ingredients from water to DNA.[108]

Hydrogen joins with oxygen to form water. Nitrogen joins with hydrogen to form ammonia. Carbon bonds with nitrogen and various other elements to create amino acids for protein synthesis and nucleic acids for genetic code.

Understanding the Electron and the EM Force

The standard model of particle physics is one of the outstanding accomplishments of twentieth century science.[109] Among seventeen total particles, it includes:

1. particles that create atoms (electrons and quarks) and
2. particles of force designed to augment and control those atoms.

We have met the bosons of the Higgs field that encode electrons and quarks with mass, and the gluons (bosons) of the strong force that lock three-quark teams into protons and neutrons.

We have recently encountered the W particle of the weak nuclear force that converts up quarks to down and triggers nuclear fusion in the stars. But we have neglected one vital particle and its

---

[107] Alison Klesman, "How Did the First Element Form after the Big Bang?" *Astronomy*, December 12, 2018, https://www.astronomy.com/science/how-did-the-first-element-form-after-the-big-bang/.

[108] Mark Fennell, *New Physical Models of the Atom, Topic D: Electron Orbits and Quantum Leaps* (self-pub., Amazon Digital Services, 2019), 140.

[109] Wouter Schmitz, *Particles, Fields and Forces: A Conceptual Guide to Quantum Field Theory and the Standard Model* (Cham, Switzerland: Springer, 2019).

force. Electrons and the electromagnetic force open the door to atoms, to molecules, and eventually to the chemistry of life.

Ninety-nine percent of the atom is surprisingly empty space. But a tiny nucleus lives at the center of the atom, delivering positive electric charge as well as most of the atom's mass. If an atom was the size of a basketball arena, the nucleus would be the tip of a shoelace.

Electrons are tiny fragments of energy with a negative electric charge. They animate our smartphones and bring heat and light to our homes. But electrons do more than flow through a current to deliver modern amenities to our lives. They also orbit every atom in our body. Electromagnetic fields course through our tissues, triggering waves in our brain and beats in our heart.[110]

Electrons orbit the atomic nucleus in "shells" like satellites orbiting Earth. The photon is the "boson" (force particle) of the electromagnetic force. Electrons can absorb the energy of a photon and jump to a higher shell. Additionally, they can emit a photon and drop to a lower shell. But the orbits of electrons are constrained by the rules of quantum mechanics.[111]

Photons of Light

As noted above, between April and September 1905, Albert Einstein submitted four papers to the *Annalen der Physik* (*Annals of Physics*). Einstein poured forth his thoughts about photons of light in a flood of insight not matched before or since by any other human being.[112]

---

[110] D. R. Westbury, "Electrical Activity of the Heart," *British Medical Journal* 4, no. 5790 (December 1971): 799–803, https://doi.org/10.1136/bmj.4.5790.799.

[111] Fennell, *New Physical Models of the Atom*, 97.

[112] Walter Isaacson, *Einstein: His Life and Universe* (New York: Simon and Schuster, 2007), 131.

Long thought to be a wave, Einstein recognized that light could also take the form of particles. A photon of light could hit a metal object and behave like a tiny BB, dislodging an electron, a phenomenon called the photoelectric effect that underlies the function of our garage door openers.

Einstein explained this behavior of photons of light with the following sentence:

"According to the assumption to be considered here, when a light ray is propagated from a point, the energy is not continuously distributed over an increasing space but consists of a finite number of energy quanta which are localized in points in space and which can be produced and absorbed only as complete units."[113]

This sentence may be difficult for us to understand, but it has been called "the most revolutionary sentence that Einstein ever wrote."[114]

Like all quantum particles, light travels as a wave, but arrives as a particle. Arriving on the surface of our retina, particles of light deliver one of God's greatest gifts, the ability to see His creation. We encounter the EM force in a variety of ways, through its electrons, photons, and magnetic fields.

A multitude of phenomena produces a flow of photons, from sunshine to wood-burning stoves to the flow of electrons through a lightbulb. Photons, Einstein tells us, also set the speed limit of the universe, circling Earth seven times per second.[115]

This chapter completes our review of the eighth and final blueprint involved in the creation of the universe (electrons and the EM force represent number eight).

---

[113] Isaacson, *Einstein: His Life and Universe*, 98.
[114] Isaacson, *Einstein: His Life and Universe*, 98.
[115] https:www.ourplnt.com. Jan 18, 2019. Ozgur Nevres: Speed of Light

Features Common to Each Blueprint

We should remember their fundamental nature:

1. Each one is described by one or more mathematical equations. The theory of relativity, an explanation of gravity, involves ten separate equations.
2. Their equations predated their arrival in the big bang package. Somehow, God converted inanimate and eternal equations into the active forces of creation.

In the remaining few chapters of section 1, we will see these independent blueprints function as a team, producing a cosmos-wide printing apparatus that fills the universe with galaxies, stars, and a wide variety of exotic space objects. We will see them create multiple generations of stars, filling the heavens with heavier elements. Finally, we will see them create planet Earth, setting the stage for the miracle of life.

We Have Met the Eighth Blueprint of Creation

$E = mc^2$

The Higgs field

Quark code

Gluons and the strong nuclear force

Quantum mechanics

Space-time

Nuclear fusion and the weak nuclear force

The electron and electromagnetic force

Car Fact

After years of enjoying the sound of a combustion engine, you have a new option to round out your fleet: an automobile fully dependent on electric power.

Supplement Seven

Thomas Edison attempted to electrify New York City with direct current, but alternating current won the day. Review the efforts of Edison and Tesla to light up the world.

CHAPTER EIGHT

Print the Cosmos

The blueprints of the universe worked in perfect harmony to deploy a printing press across the heavens. That printing power created untold numbers of galaxies, stars, planets, and various exotic space objects, producing the wonders of the cosmos.

In the 1920s, scientists thought our Milky Way galaxy was the entire universe. But Edwin Hubble peered through his telescope and saw faint blurs in the background of space. Soon he realized that these "island universes" were independent galaxies. Even more, these distant galaxies were speeding away from each other like race cars, suggesting that the universe was expanding. Rewind the tape, and all this content once existed in a single atom. Hubble joined Lemaître in embracing the big bang concept.[116]

The Hubble Space Telescope,[117] now joined by the even more powerful James Webb Space Telescope,[118] has given us a window into the grandeur and mystery of space, capturing light that began its journey billions of years in the past.

---

[116] Simon Singh, *Big Bang: The Origin of the Universe* (New York: Harper Perennial, 2005).

[117] Jim Bell, *Hubble Legacy: 30 Years of Discoveries and Images* (New York: Sterling 2020).

[118] James Webb Space Telescope (website), NASA, last modified July 24, 2024, https://www.jwst.nasa.gov/.

The Two Clear Objectives of the Printing Program

1. Create a star-planet combination ideally suited for life.
2. Surround that star-planet combination with an enormous scaffolding of galaxies and stars, dark matter, and dark energy, all blended in a perfect recipe to give Earth a long and stable ride.

The Creative Power of Impact

In the early days of our solar system, a Mars-sized object plowed into newborn Earth.[119] As we will see in an upcoming chapter, the collision had far-reaching consequences, creating the moon and converting our protoplanet into an incubator for life.

The creation of the moon was only the first collision to shape the planet. Sixty-five million years ago, a space object called an asteroid that was six to seven miles in diameter plowed into the Yucatán Basin, releasing a billion times more energy than the atomic bombs that destroyed Hiroshima and Nagasaki. An earthquake followed—as did a thousand-foot-high tsunami and six hundred mph winds.[120]

The event blocked out the sun and ended the reign of the dinosaurs.

---

[119] Robin M. Canup and Erik Asphaug, "Origin of the Moon in a Giant Impact Near the End of the Earth's Formation," *Nature* 412 (September 2001): 708–712, https://doi.org/10.1038/35089010.

[120] Riley Black, *The Last Days of the Dinosaurs: An Asteroid, Extinction, and the Beginning of Our World* (New York: St. Martin's Press, 2022).

> **Question:**
> *Dinosaurs ruled Earth for millions of years, but mammals made better parents and may have had more promising potential to develop larger brains. Are we surprised that mammals would survive extinction and become a dominant life-form?*

Deploying the Space-Time Printer

The printing platform assembled by the eight blueprints has filled the universe with extraordinary objects:

1. Hundreds of billions of galaxies[121]
2. Two hundred billion trillion stars[122]
3. Seven hundred quintillion planets[123]
4. Two thousand neutron stars in the milky way[124]
5. One hundred billion supermassive black holes[125]
6. One to two million asteroids in a zone between Mars and Jupiter[126]
7. So far, 3,957 comets have been found (as of July 29, 2024[127]) and maybe trillions that remain unknown

---

[121] Ethan Siegel, "This Is How We Know There Are Two Trillion Galaxies in The Universe," *Forbes*, October 18, 2018, https://www.forbes.com/sites/startswithabang/2018/10/18/this-is-how-we-know-there-are-two-trillion-galaxies-in-the-universe/.

[122] https://bigthink.com/hard-science/how-many-stars-are-in-the-universe/ Oct 4, 2021

[123] Tibi Pulu, November 28, 2019,

[124] htpps://www.astronomy.com/science/what-are-neutron-stars-the-cosmic-gold-mines-explained/

[125] Kormendy, John; Richstone, Douglas (1995), "Inward Bound—The Search For Supermassive Black Holes In Galactic Nuclei", *Annual Review of Astronomy and Astrophysics*, 33: 581.

[126] https://en.wikipedia.org › wiki › Asteroid

[127] "Solar System Objects," Solar System Dynamics, NASA Jet Propulsion Laboratory, accessed July 29, 2024, https://ssd.jpl.nasa.gov/.

No Wasted Space

Lest these enormous numbers intimidate readers, we should keep one thought in mind: we needed a universe of immense scale, filled with an abundance of printed products, to provide a stable scaffolding for the development of life on Earth. As the television personality and physicist Carl Sagan is credited with saying, "The universe is a pretty big place. If it's just us, it seems like an awful waste of space."[128]

With all due respect to Professor Sagan, it took every bit of a universe almost infinite in size, filled with vast numbers of mass-encoded objects, to give Earth a stable ride and the time required to produce a cornucopia of living creatures. Recognizing that cosmic scaffolding was vital for the development of life on Earth, we can appreciate that the universe has no wasted space.

Hopefully, the take-home point of this section is clear. Eight unique programs (sets of equations or blueprints) conspired perfectly to make our universe a cosmic-size printing apparatus. Remove any blueprint, and there would be no printed products. Without all eight working as a team, there would be no galaxies, stars, black holes, comets, or asteroids. There would be no planets or living creatures. There would be no human footprints on Earth.

Little wonder the King David looked at the nighttime sky and celebrated the work of God's fingers. "Lord, our Lord, how majestic is your name in all the earth!" and "When I consider your heavens, the work of your fingers, the moon and the stars which you have set in place, what is mankind that you are mindful of them, human beings that you care for them?"[129]

---

[128]   Carl Sagan, *Contact* (New York: Simon and Schuster, 1985).
[129]   Psalm 8:9 (NIV).

More to the Story: Discovering the Universe

From the earliest moments of human history, humanity has studied the heavens. Archaeological records show that astronomy was one of the first natural sciences embraced by early civilizations. Ancient astronomers closely monitored the positions of celestial bodies, assigning signs and warnings to changes in the sky.[130]

Imagine the excitement of ancient sky watchers when comets blazed across the heavens. Snowballs of frozen gas, rock, and dust orbiting the sun, comets can reach the size of a small town. Edmund Halley witnessed a spectacular comet in 1682 and asked his friend Isaac Newton to provide a gravitational analysis of its flight.[131]

Newton responded with unwelcome news: he had solved the gravitational equations of the comet many years earlier but had lost his notes. Halley convinced him to rework his calculations and paid for their publication.

Using Newton's analysis, Halley realized that the comets of 1531, 1607, and 1682 were the same comet—a comet that would return every seventy-six years. The comet, he predicted, would return in 1759. Haley died in 1742, but his comet returned on schedule, not only in 1759 but also in 1835, 1910, and 1986. Stay tuned for Haley's return in 2061.[132]

Over the course of more than two thousand years, different cultures have documented their sightings of Halley's comet. It was

---

[130] James Evans, *The History and Practice of Ancient Astronomy* (Oxford: Oxford University Press, 1998).

[131] John Nobel Wilford, "Sir Edmund Halley: Orbiting Forever in Newton's Shadow," *New York Times*, October 29, 1985, https://www.nytimes.com/1985/10/29/science/sir-edmund-halley-orbiting-forever-in-newton-s-shadow.html.

[132] "1P/Halley," Science, NASA, last updated January 25, 2024, https://science.nasa.gov/solar-system/comets/1p-halley/.

noted in Chinese records in 240 BC[133] and on Babylonian tablets in 164 BC.[134]Perhaps its most famous sighting occurred 1n 1066 AD at the time of the Norman invasion of England that resulted in the death of King Harold II and the takeover of England by the Norman elite. This dramatic moment in European history was famously recorded for posterity by the seamstresses of the Bayeaux Tapestry, 230 feet long and on display at the Bayeux Museum in Normandy France.[135]

At the end of an all-day bloody battle, King Harold II received a fatal blow—shot in the eye with an arrow.

Edwin Hubble and His Telescope

While meteors and asteroids have visited Earth from outer space since the beginning of recorded history, it is only in the past century that science has returned the favor. Using the most powerful telescope ever engineered by human hands at that time, Edwin Hubble photographed distant lights in the sky.[136] As noted earlier in this chapter, Hubble and his telescope soon proved that the universe was much grander than a single galaxy, the Milky Way.

By 1922, Hubble had undeniable proof. The objects of his fascination were vast collections of stars. He called them "island universes." Vaster than humanity had ever imagined, the universe

[133] Gary W. Kronk, *Cometography, Volume 1: Ancient – 1799* (Cambridge: Cambridge University Press, 1999), 14.
[134] F. R. Stephenson, K. K. C. Yau, and H. Hunger, "Records of Halley's Comet on Babylonian Tablets," *Nature* 314 (April 1985): 587–592, https://doi.org/10.1038/314587a0.
[135] Donald K. Yeomans, "Great Comets in History," NASA Jet Propulsion Laboratory, April 2007, https://ssd.jpl.nasa.gov/sb/great_comets.html.
[136] Gale E. Christianson, *Edwin Hubble: Mariner of the Nebulae* (Boca Raton, Florida: CRC Press, 1997).

stretched far beyond the Milky Way. Humankind, we can say without exaggeration, "discovered the universe" in 1925.

Hubble soon made an even more stunning discovery. Using the spectra of light radiating from these distant islands, he classified galaxies by their "red shift," indicating how fast they were receding from view. Closer galaxies were moving slowly, but more distant galaxies were speeding up. A galaxy ten million light-years from the Milky Way was traveling twice as fast as a galaxy five million light-years away.[137]

Soon, the accumulating data formed an undeniable picture. Space-time itself was expanding. Galaxies were moving further and further apart, as though they were surfing "on space-time's wave."

Physicist James Jeans provided a clear explanation: "This looks as though the whole universe were uniformly expanding, like the surface of a balloon while it is being inflated…. The properties of the space in which they exit compel them to scatter."[138]

THE UNIVERSE IS EXPANDING

As the universe expands like inflating balloons, galaxies grow further and further apart.

137   Edwin Hubble, "A Relation Between Distance and Radial Velocity Among Extra-Galactic Nebulae," *Proceedings of the National Academy of Sciences of the United States of America* 15, no. 3 (March 1929): 168–173, https://www.pnas.org/doi/pdf/10.1073/pnas.15.3.168.
138   Marcia Bartusiak, *The Day We Found the Universe* (New York: Pantheon Books, 2009), 67.

The Hubble and James Webb space telescopes have delivered thousands of jaw-dropping images to Earth, helping astronomers determine the universe's age and discover planets, quasars, and much more.[139]

God's printing, it seems, has been more prolific than we can imagine.

Car Fact

Halley's comet might be considered the first fully autonomous vessel known by humans. It circumnavigates a vast territory of space and returns to a precise location in the heavens every seventy-six years.

Supplement Eight

In addition to two space telescopes, scientists have deployed a variety of sophisticated devices to explore the cosmos. Human beings are still discovering the universe.

---

[139] Bell, *Hubble Legacy.*

CHAPTER NINE

Farm the Stars

Our universe has given birth to multiple generations of stars. As element-rich stars go supernova, they share their contents with future stars. Carbon takes the perfect quantum leap to double its production in these shining bodies, confirming the divine intention that life should inhabit Earth.

Initial stars were element poor, so massive that their lifespan was short.[140] But each generation of stars became more chemical rich as fusion converted helium into nitrogen, carbon, oxygen, and heavier elements.

---

[140]   Emma Chapman, *First Light: Switching on Stars at the Dawn of Time* (London, UK: Bloomsbury Sigma, 2022), 15.

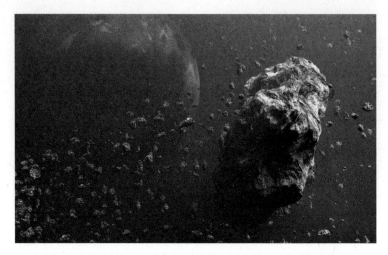

The printing power of space-time in collaboration with nuclear fusion has filled the universe with exotic space objects. Here an asteroid passes a giant red star.

Main sequence stars, like our sun, fuse hydrogen into helium in their core.[141] When they exhaust the core hydrogen, they rapidly swell into red giants and fuse hydrogen in an outer shell. When red giants exhaust their source of hydrogen and fusion finally stops, they collapse in minutes and explode as a supernova.[142] The death of these older giants fertilized the heavens and gave newborn stars a head start.

Double the Production of Carbon

God fertilized the heavens and rotated His stars, taking particular care to produce adequate carbon for life. Ninety-two elements occur

---

[141] Leon Golub, *Nearest Star: The Surprising Science of Our Sun* (Cambridge: Harvard University Press, 2001),30.

[142] Laurence A. Marschall, *The Supernova Story* (Princeton: Princeton University Press, 1988).

naturally on Earth, but only eleven in more-than-trace quantities. Life utilizes six elements, most importantly carbon. Carbon represents 18 percent of our biomass.[143] We needed abundant carbon to produce the constituents of life, including DNA and proteins.

The fusion of two helium-4 atoms should logically produce beryllium-8, but God had a different plan. In the center of the stars, carbon takes a perfect quantum leap to form a three-way entanglement with helium and beryllium. Carbon, not beryllium, emerges from the process. Thanks to its quantum leap in the stars, carbon is the sixth most common element in the universe. Beryllium ranks sixty-first.[144]

Creating Our Solar System

Following two to three cycles of star birth and death, a new solar system with a just-right sun and eight trailing planets emerged from ancestral dust.

Question:
In 1953, the British cosmologist Fred Hoyle predicted carbon-12 would possess quantum features that would maximize its production in the stars. His prediction was soon confirmed by laboratory testing. Is it coincidental that carbon-12 dwarfs the abundance of beryllium-8, its precursor in the stars?

---

[143] Franklin M. Harold, *On Life: Cells, Genes, and the Evolution of Complexity* (Oxford, UK: Oxford University Press, 2022), 3.

[144] Wikipedia, s.v. "Abundance of the Chemical Elements," last modified July 24, 2024, 16:09, https://en.wikipedia.org/wiki/Abundance_of_the_ chemical_elements.

The Dawn of the Universe

The newborn universe was too hot and energetic to allow the formation of stars. But as the universe expanded and cooled, it experienced a cosmic dawn.[145] Mammoth stars erupted into luminous balls of heat and light as gravity conspired with the strong and weak nuclear forces to light up the darkness of space.

The earliest stars were composed entirely of primordial gas made in the first few minutes of creation. They contained hydrogen, helium, and lesser amounts of lithium and beryllium, a steel gray, metal-like aluminum. But each generation of stars produced more quantities of heavier elements. Most stars today contain iron, nickel, carbon, and oxygen, in addition to lighter elements.[146]

Astronomers today classify the most abundant stars as "main sequence."[147] These can vary in size, brightness, and mass, but they all conduct the same nuclear fusion in their cores. They fuse hydrogen into helium, releasing vast quantities of energy and pushing back against the force of gravity. Gravity compresses the star inward. Nuclear fusion pushes outward. These forces balance each other out, and the star remains a sphere.

---

[145]   Chapman, First Light.
[146]   Golub, *Nearest Star*, 70.
[147]   J. J. Eldridge and Christopher A. Tout, *The Structure and Evolution of Stars* (Hackensack, NJ: World Scientific, 2018), 25.

There are a variety of star types, each with their own life cycle. Here a sun-like star consumes its fuel, morphs into a giant red star before going supernova and dying in a stunning explosion. Dying stars fill space with a variety of elements created by stellar fusion.

When a mainline star exhausts the fuel in its core, it begins to burn hydrogen in its outer shells. The star cools, expands tenfold, and becomes a red giant.[148] Eventually, a red giant runs out of hydrogen and ejects its outer shell. A pure iron core collapses in seconds and triggers a staggering explosion called a supernova. Supernovas produce a shock wave, creating the heaviest elements.

The Fate of a Star

The mass of the star determines its final fate. Smaller stars become white dwarfs. If one white dwarf collides with another white dwarf or pulls too much matter from another star, it can explode, creating a Type Ia supernova.[149]

---

[148] Eldridge and Tout, *Structure and Evolution*, 24.
[149] Marschall, *Supernova Story*, 120.

When a more massive star runs out of fuel and collapses, it creates a Type II supernova.[150] An object more than one million times the size of Earth collapses in fifteen seconds. For a few days, the supernova is the brightest object in its galaxy, with a luminosity exceeding a trillion suns.

During the supernova explosion, the star dispenses its chemical riches into an expanding cloud of nebular gas, fertilizing the nursery for future stars.

Dying stars with a mass twenty-five times greater than the sun have a more exotic fate: they can collapse into a neutron star, a small ball composed of tightly packed neutrons.[151] With a diameter of only ten to twenty miles, one teaspoon of a neutron star can weigh a billion tons!

The most massive stars can collapse into black holes with a gravitational field so intense, not even light can escape its embrace.

The Master Star Plan

Our universe contains 125 billion galaxies. Each galaxy has at least one hundred million stars. Readers good at math can multiply 125 billion times one hundred million and appreciate the enormity of star production. In shorthand, the universe contains more than one million trillion stars.

Star Farming

God commissioned star farming. Hydrogen gas provided the fuel. Short-lived supergiant stars fertilized the fields, creating a bumper crop of main sequence stars. Over time, the blueprints of the

---

150   Marschall, *Supernova Story*, 120.
151   Marschall, *Supernova Story*, 26.

universe set the stage for a solar system, the sun and eight planets, and the appearance of life on Earth.

Car Fact

Ancient stars were a perfect example of planned obsolescence. Just as new cars deliver valuable features that consign older models to the dustbin of history, stars in the universe steadily acquire enhanced chemical capabilities. Our sun is the latest model.

Supplement Nine

NASA sponsored "SETI@home," incorporating the computers of tens of thousands of volunteers to listen for life in the universe. The result is well known: no one phoned home. SETI 2.0 is now underway.[152]

So far, all is silent.

---

[152] The SETI Institute launched SETI 2.0 in 1999.

CHAPTER TEN

Earth

Earth is a spectacular planet, tailor-made to be a platform for life.

Shortly after the emergence of our solar system, a dramatic event reshaped Earth. A space object the size of Mars (half the size of our planet) plowed into proto-Earth at a sixty-degree angle and a speed of nine thousand miles per hour.[153] The result of this collision was far reaching:

The merger created a new and more massive Earth with a just-right gravitational field, strong enough to attract and hold a life-supporting atmosphere.

Ejected debris created a perfect moon, one of the largest in our solar system, with a tilt that moderates the wobble on our axis and promotes a stable climate.[154] The moon also controls our ocean tides.

The impact melted the Earth's iron, creating a molten core, magnetic fields, and a just-right crust with tectonic plates.[155] Each

---

[153]   Robin M. Canup and Erik Asphaug, "Origin of the Moon in a Giant Impact Near the End of the Earth's Formation," *Nature* 412 (September 2001): 708–712, https://doi.org/10.1038/35089010.

[154]   Kaveh Pahlevan and Alessandro Morbidelli, "Collisionless Encounters and the Origin of the Lunar Inclination," *Nature* 527 (November 2015): 492–494.

[155]   Robert M. Hazen, *The Story of Earth: The First 4.5 Billion Years, from Stardust to Living Planet* (New York: Viking, 2012), 27.

feature of the newly formed Earth and its moon was essential for the creation of life.

Experts consider Earth "rare,"[156] suggesting it is the only life-supporting planet in the cosmos. Others describe it as "privileged,"[157] perched in a perfect position on a spiral arm of its galaxy to give humanity a front-row view of the cosmos.

God made Earth a platform for life. He also assured that its human inhabitants could study and celebrate creation!

Scientists have searched for alien life since the 1920s. So far, nothing has been found. No other planet, it seems, has a perfect mass at a just-right distance from their mother star, with an ideal moon, tectonic plates, magnetic fields, and an atmosphere supportive of life.

Section 2 will address a related question: how many planets beside Earth are home to DNA?

Question:
The impact that created the moon created magnetic fields, tectonic plates, enough mass to retain a perfect atmosphere, moderate climate, and other life-friendly features. Did planet Earth simply get lucky?

---

[156] Peter D. Ward and Donald Brownlee, *Rare Earth: Why Complex Life is Uncommon in the Universe* (New York: Copernicus, 2000).
[157] Guillermo Gonzalez and Jay W. Richards, *The Privileged Planet: How Our Place in the Cosmos Is Designed for Discovery* (Washington DC: Regnery Publishing, 2004).

Earth is an Extraordinary Planet

Earth is a remarkable planet possessing multiple life-friendly features. This image reveals its four major layers, an inner and outer core, a mantle, and crust.

Earth has liquid water, plate tectonics, and an atmosphere that shelters it from the worst of the sun's rays. It rotates a perfect distance from the sun. A planet closer to the sun would receive too much energy and lose its atmosphere, while a planet too far out would quickly freeze.[158] It takes about eight minutes for sunlight to reach our planet.

Plate tectonics working over geological time are a vital piece of the puzzle.[159] The slip-sliding movements of the Earth's crust have created the planet's towering mountain ranges and vast ocean depths. That same sliding crust drives a carbon-silicate cycle that regulates carbon levels in the atmosphere and adjusts surface

---

[158] Ward and Brownlee, *Rare Earth*, 17.
[159] Ward and Brownlee, *Rare Earth*, 110.

temperatures.[160] The narrow range of our surface temperature maintains an extraordinary amount of liquid water. Water lubricates plate tectonics, and plate tectonics slowly build mountains. Without liquid water, the Earth would be geologically dead.[161]

Rare Planet

Earth is also a just-right size. Smaller, and it would lose its atmosphere. Larger, and it might retain poisonous gases in its atmosphere and experience temperatures too extreme for life.[162] Our atmosphere and its ozone layer protect us from harmful radiation while providing proper sunlight for flowers and plants.[163]

Earth enjoys another favorable feature; it has a big brother called Jupiter circling in the outer solar system, deflecting most incoming asteroids and space debris.[164]

Earth's axis of rotation around the sun tilts at 23.4 degrees along an imaginary plane slicing through the Earth's equator.[165] While people living on the equator experience the same weather year round, those living further north and south experience seasons produced by the Earth's tilt. The Northern Hemisphere bends

---

[160] Joshua Krissansen-Totton and David C. Catling, "A Coupled Carbon-Silicon Cycle Model over Earth History: Reverse Weathering as a Possible Explanation of a Warm mid-Proterozoic Climate," *Earth and Planetary Science Letters* 537 (May 2020): article 116181, https://doi.org/10.1016/j.epsl.2020.116181.

[161] Ward and Brownlee, *Rare Earth*, 194.

[162] Frederick K. Lutgens, Edward J. Tarbuck, and Redina L. Herman, *The Atmosphere: An Introduction to Meteorology*, 6th ed. (Hoboken, NJ: Prentice Hall, 1995).

[163] Robert M. Hazen, *The Story of Earth: The First 4.5 Billion Years, from Stardust to Living Planet* (New York: Viking, 2012), 158.

[164] Ward and Brownlee, *Rare Earth*, 235.

[165] Charles Q. Choi and Ailsa Harvey, "Planet Earth: Everything You Need to Know," Space.com, last updated April 13, 2023, https://www.space.com/54-earth-history-composition-and-atmosphere.html.

toward the sun during part of the year, while the Southern Hemisphere tilts away.

With the sun higher in the sky, solar heating is more intense in the north, producing summer. Six months later, the situation reverses. When spring and fall begin, both hemispheres receive equal amounts of heat from the sun.

Where Else Would You Like to Live?

You should think twice before making Mars your second home. It has a minimal atmosphere and excessive radiation from the sun.[166] Would you prefer Venus? Venus has no surface water, no plate tectonics, and the average temperature exceeds four hundred degrees.[167] How about all those planets scientists discover so frequently? Many are "hot Jupiters," too deprived of oxygen to support the origin of life.[168]

Other rocky planets orbit red dwarf stars producing too little heat and light. Forced to snuggle close to these red dwarfs in search of a "habitable zone," these planets are pounded by extreme radiation. A recent analysis of conditions on the "Earth-like" planet Proxima b, just four light-years away, reported ominous findings: "seven-second death rays" of radiation repeatedly "thrash" the planet.[169]

---

[166] Gonzalez and Richards, *Privileged Planet,* 84.
[167] Gonzalez and Richards, *Privileged Planet,* 192.
[168] John Wenz, "What Astronomers Can Learn from Hot Jupiters, the Scorching Giant Planets of the Galaxy," *Smithsonian Magazine,* October 11, 2019, https://www.smithsonianmag.com/science-nature/what-astronomers-can-learn-hot-jupiters-scorching-giant-planets-galaxy-180973320/.
[169] Jamie Carter, "Our Neighbors are Probably Dead. The Closest Earth-Like Planet to Us is Being Thrashed by 7-Second 'Death Rays'" *Forbes,* last updated December 10, 2021, https://www.forbes.com/sites/jamiecartereurope/2021/05/04/our-neighbors-are-dead-the-closest-earth-like-planet-to-us-is-being-thrashed-by-7-second-death-rays/.

Are We Alone?

Scientists agree there must be billions of planets in our galaxy. Undoubtedly, they assert, some must have fostered the development of advanced forms of life. Enrico Fermi asked the obvious question: if that is the case, why haven't they gotten in touch?[170]

In a well-received book entitled *Rare Earth: Why Complex Life is Uncommon in the Universe*, two thoughtful scientists argue that intelligent life is rare, so rare that it may be an exclusive feature of Earth.[171] The conditions for the evolution of intelligent life, from a just-right moon to plate tectonics to a just-right distance from a just-right star (not to mention the availability of DNA), may be so unusual that we may be alone. Conditions in much of the universe may be fundamentally hostile to life.

We end this tour of cosmic blueprints with the creation of a "rare" and "privileged" Earth awaiting the first sign of life.

Car Fact

Modern automobiles are a marvel of style and engineering. They come equipped with seat belts, air bags, and variable cruise control, all designed to protect passengers' lives.

Who designed spaceship Earth to be friendly to life as well?

---

170   Gonzalez and Richards, *Privileged Planet,* 276.
171   Ward and Brownlee, *Rare Earth,* 235.

Supplement Ten

Learn how an ancient Greek named Eratosthenes measured the circumference of Earth with a result that matches modern GPS calculations.

Also review the voyage of Ferdinand Magellan, as his round-the-world voyage sacrificed ships and men to prove the Earth is round.

The Divine Universe

The big bang delivered near-infinite quantities of heat and light. But to what end, if energy failed to congeal into particles that would create electrons and quarks for atoms, gluons and W particles for the strong and weak nuclear forces, photons for electromagnetism and vision, and bosons for the Higgs field?

Quarks congealed from big bang light. But to what purpose, if a Higgs field failed to encode them with mass, or a strong force failed to lock them inside protons and neutrons?

Quarks came in different flavors, including up and down, with curious fractions of electric charge. But to what creative end, if one three-quark team failed to produce a positive proton, while another failed to produce a charge zero (neutral) neutron?

Nuclear fusion in the stars would be a futile dream without the warping of space-time and gravity's squeeze. Nuclear fusion would also be impossible without the W particle of the weak nuclear force converting up quarks to down, producing copious quantities of deuterium and tritium.

What would nuclear fusion accomplish if the release of mass is multiplied by an insignificant number rather than by the unimaginably large factor of c^2 (299,792,458 meters per second)?

But to what purpose, if the elements are forged in the stars without the electrons to complete atoms and form the molecular foundation for life?

All that star printing across the cosmos would be a gargantuan waste of energy and time without the creation of a rare and special Earth capable of supporting life and the development of minds.

How Many "What Ifs?"

How many "what ifs" does it take, how many interactive component parts, how many potential dead ends rescued by just-so fields and forces before we celebrate God's creative power?

The prophets of "scientism" have failed to place science in its proper perspective.[172]

Arguments for the Existence of God

Arguments for the existence of God are as old as philosophy itself.[173] Plato, the world's leading philosopher, laid the foundation for this argument in 400 BC. Plato believed in "ideal Forms," timeless, perfect, and abstract concepts that occupy a "realm of Forms," a reservoir of blueprints for the assembly of our world.[174]

Christian philosophers perfected Plato's concept. The first-century Christian apologist Justin Martyr viewed "the cosmos" as "permeated by seeds of the divine word (logos)."[175]

---

[172] John Polkinghorne, *Quarks, Chaos and Christianity: Questions to Science and Religion* (Chestnut Ridge, Crossroad, 1994), 10.

[173] Jerry L. Walls and Trent Dougherty, *Two Dozen (or so) Arguments for God: The Plantinga Project* (Oxford, UK: Oxford University Press, 2018).

[174] Plato, *Great Dialogues of Plato*, trans. W. H. D. Rouse (New York: Signet, 2015).

[175] Saint Justin Martyr, *The Writings of Justin Martyr*, ed. Alexander Roberts D. D., James Donaldson L. L. D., and Paul A. Boer Sr. (self-pub, CreateSpace, 2014).

Two centuries later, Augustine of Hippo articulated a widespread view of early Christians: the development of the cosmos and everything within it are "governed by fundamental laws which reflect the will of their creator."[176] Those laws do not exist out of necessity. God created them because He wanted to and out of goodness.

Thomas Aquinas

Theologian and philosopher Thomas Aquinas wrote five proofs for the existence of God, arguing that "All things have an order or arrangement that leads them to a particular goal. Because the order of the universe cannot be the result of chance, design and purpose must be at work." Recognizing divine intelligence at work in the universe, Aquinas was a strong advocate for "Natural Theology," celebrating God's work in the natural order.

---

[176] Saint Augustine, *The City of God*, trans. Marcus Dods, D. D. (New York: Modern Library, 1993).

A thousand years later, in his *Summa Theologica*, Thomas Aquinas provided five arguments for the existence of God.[177]

God is the:

- Unmoved Mover
- The First Cause
- The Necessary Being
- The Absolute Being
- The Grand Designer

In his "fifth way," Aquinas argued for an intelligent creator "based on perceived evidence of deliberate design in the natural or physical world." These arguments provide the basis for natural theology, celebrating God through His scientific works. Science can tease apart creation and recognize its ingenuity and purpose.

As Aquinas stated, "We see that things which lack intelligence, such as natural bodies, act for an end, and this is evident from their acting always, or nearly always, in the same way, so as to obtain the best result. Hence it is plain that not fortuitously, but designedly, do they achieve their end.… Therefore, some intelligent being exists by whom all natural things are directed to their end; and this being we call God."[178]

The doctrine of incarnation declared by the first Council of Nicaea in AD 325 summarized creation for the Christian.[179] Plato's "Form of the Good" took on human form as God incarnate lived

---

[177]   Gilbert Keith Chesterton, *St. Thomas Aquinas: The Dumb Ox* (New York: Doubleday Image Books, 1998).

[178]   Thomas Aquinas, *Summa Theologica*, trans. Fathers of the English Dominican Province (Notre Dame, IN: Christian Classics,1948). 43.

[179]   David E. Henderson and Frank Kirkpatrick, *Constantine and the Council of Nicaea: Defining Orthodoxy and Heresy in Christianity, 325 CE,* (Chapel Hill: University of North Carolina Press, 2016).

on His "rare" and "privileged" planet and confirmed God's power over death.

Christ gave us undimmed access to the truth. Rather than engaging in mystical meditation, we can look at Jesus. "For in him dwelleth all the fulness of the Godhead bodily."[180]

Unfathomable Mind

Should you be skeptical that the blueprints that created our universe confirm the existence of God? Surely each one belongs to the domain of science and has little to do with faith. While each one is a marvel of design, the ultimate evidence for superintelligent planning comes from their aggregate actions.

- Without the Higgs mechanism, the eruption of the big bang might have blazed brightly on its own, producing a magnificent light show. But no protons would have formed, no atoms or molecules would exist, and no human would exist to care about a Higgs equation.
- The Higgs field could have filled the universe as an empty sea of lonely bosons with no particles to encode with mass. The universe would have been eternally dark and lifeless.
- Space-time might have undulated across an expanding universe unaccompanied by quarks or Higgs bosons. The universe would be a cosmic wasteland.

Each component of God's printing press is a distinct creative impulse. But it took their collaboration to accomplish divine intention. Without all eight programs working in perfect concert, there would be no universe to print. Does the system seem like

---

[180] Colossians 2:9 (NIV).

the random invention of a mindless nature? Does the process seem accidental? No. The prowess of these creative blueprints confirms a singular worldview...

God is Real and the Reason We Exist

While we may be able to identify the blueprints of creation, it is something else to understand how God ignited their creative interaction. But we can say a few things for sure:

- No accident of nature cared that a proton could become a neutron by converting a up quark to a down, creating deuterium and tritium in the stars.
- No accident of nature assured that unrelated nuclear forces, the strong and the weak, would conspire with gravity to trigger the shining of the sun.
- No accident of nature designed the table of elements and the supernovae that fertilized our sun and prepared the raw material for Earth and its myriad of life-forms.

Planned Before Time

Let us focus on a startling example of God's planning in advance. In the first few minutes of creation, up and down quarks created sufficient protons to fuel every future star. At the same time, W particles of the weak nuclear force joined protons (hydrogen clouds) in an epic journey to the first star.

Computer models suggest the first star was born 150–200 million years after the big bang.[181] After millions of years in

---

[181] Emma Chapman, *First Light: Switching on Stars at the Dawn of Time* (London, UK: Bloomsbury Sigma, 2022).

limbo, the weak force began to execute its mission, converting up quarks into down and producing fuel for the first ancient star. The mechanism that converts balls of hydrogen gas into nuclear reactors lay dormant for millions of years until God declared once again: let there be light!

The universe was born with up and down quarks, with strong and weak nuclear forces, and with space-time and gravity destined to create the glory of the nighttime sky. In the fullness of time, those blueprints accomplished God's intention.

Loving God with Your Mind

Most of us lack the mathematical skills to decipher the equations of creation. But we can understand their fundamental mission. As John Lennox reminds us,[182] Jesus's first commandment was to "Love the Lord your God with all your heart and with all your soul and with all your mind and with all your strength (Mark 12:30)." Notice the inclusion of "mind" in this commandment.

Lennox also makes a connection between God and science: "On the doorway of the famous Cavendish Physics Laboratory in Cambridge, Sir James Clerk Maxwell had the words of Psalm 111 carved above the door: 'The works of the Lord are great, sought out of all them that have pleasure therein (Psalm 111:2, KJV [King James Version])."[183]

---

[182]   John C. Lennox, *Can Science Explain Everything?* (Charlotte: The Good Book Company, 2019), 43.
[183]   Lennox, *Can Science Explain Everything?*, 44.

This image reminds us once again that multiple designer blueprints produced the unimaginable universe we live in today. Those mathematically based components acted seamlessly together to create a magnificent cosmos "as interesting as possible."

What have we encountered in the hallowed halls of science? We have met an ensemble of creative forces, interacting with symphonic precision to print out a universe that challenges our comprehension. Why have scientists failed to acknowledge the divine implications of these blueprints? Are they hoping to discover a "theory of everything" that wraps the blueprints into a tidy package and makes the concept of a creator obsolete?

A Theory of Everything?

If a theory of everything exists, it must embrace Newton's laws, Maxwell's electromagnetic mathematics, and Einstein's gravitational equations. Such a theory would have to incorporate symmetries that determine which particle interactions can occur and which cannot. It must be compatible with four dimensions, including space and time.

John Barrow, British astrophysicist who received the 2006 Templeton Prize for Progress Toward Research or Discoveries About Spiritual Realities, and author of *Theories of Everything: The Quest for Ultimate Explanation* is not impressed: "There is no formula that can deliver all truth, all harmony, all simplicity. No Theory of Everything can ever provide total insight."[184]

Our modern military might provide a helpful metaphor. Can we combine air power, ground troops, tanks, missiles, drones, and nuclear armaments into a single algorithm?

Our universe shares similar features: an armada of energy, fields, forces, particles, space-time, and quantum mystery weaves the fabric of our world. Stephen Hawking expressed his surprise and asked the obvious questions: "What is it that breathes fire into the equations and makes a universe for them to describe... Why does the universe go to all the bother of existing?"[185]

Einstein's Search for a United Theory

In his later life, Einstein hoped to merge his theory of relativity with the theory of electromagnetic fields. His efforts were fruitless.

But the search for a unified theory continues as scientists attempt to merge the theory of relativity with quantum mechanics

---

[184]　John D. Barrow, *Theories of Everything: The Quest for Ultimate Explanation* (Oxford, UK: Oxford University Press, 1991).
[185]　Stephen W. Hawking, *A Brief History of Time* (New York: Bantam Books, 1998), 190.

to develop a quantum theory of gravity.[186] Fundamental particles display quantum features. But gravity and the universe writ large involves the language of classical physics. A quantum theory of gravity would allow quantum fluctuations of space-time to play an important role at the center of a black hole. Is a theory of quantum gravity on the horizon? Perhaps even suggesting the bulk of the universe is holographic?[187]

We should revisit the broad attributes of quantum theory. Particles arrive in the classical world as probability equations: a particle explores its future options as a superposition, a tableau of possibilities. It may reach its destination along this path or another or even along a near-infinite assortment of paths. Can we comprehend this behavior, or are these particles the functional equivalent of extraterrestrials arriving from another world?

A theory of quantum gravity might explain space-time at the smallest scale. But no theory can explain quantum properties themselves.

Nor could quantum gravity explain energy-mass equivalence ($E = mc^2$), a universe-wide mass-encoding field, or quarks imbued with fine-tuned fractional charge.

What Sort of God?

Paul Davies summarizes the problem: "It seems to me that if one perseveres with the principle of sufficient reason and demands a

[186] C. Marletto and V. Vedral, "Gravitationally Induced Entanglement between Two Massive Particles is Sufficient Evidence of Quantum Effects in Gravity," *Physical Review Letters* 119, no. 24 (December 2017): article 240402, https://doi.org/10.1103/PhysRevLett.119.240402.

[187] Anil Ananthaswamy, "Is Our Universe a Hologram? Physicists Debate Famous Idea on its 25th Anniversary," *Scientific American*, November 30, 2022, https://www.scientificamerican.com/article/is-our-universe-a-hologram-physicists-debate-famous-idea-on-its-25th-anniversary/.

rational explanation for nature, then we have no choice but to seek that explanation in something beyond or outside the physical world—in something metaphysical." He continues: "If one is prepared to go along with the idea that the universe does not exist reasonlessly, and if for convenience we label the reason God...in what sense might God be said to be responsible for the laws of physics (and other contingent features of the world)?... So what sort of God would this be?"[188]

Theologians have provided consistent answers to this question. Such a God would be omnipotent and omniscient, creating the universe ex nihilo and guiding its unfolding in a perfectly rational way. How might an omniscient and omnipotent God commission the laws of physics? The existence of a mathematical landscape might provide one "causal joint."[189] He may have selected equations from that library, somehow "set them on fire," and dispatched them to His purpose.

Should we doubt the existence of such an all-encompassing mathematical landscape occupying another dimension? Great mathematicians believe they visit such a library to unlock the secrets of the universe.

Libraries for Math and Life

Even more astounding, scientists with no religious agenda have used supercomputers to describe a similar library for life.[190] Would our unfathomable Creator establish a library for the cosmos and

[188] Paul Davies, *The Mind of God: The Scientific Basis for a Rational World* (New York: Simon and Schuster, 1993), 239.

[189] Roger Penrose, *The Road to Reality: A Complete Guide to the Laws of the Universe* (New York: Vintage Books, 2004).

[190] Andreas Wagner, *Arrival of the Fittest: Solving Evolution's Greatest Puzzle* (New York: Current, 2014).

leave the journey of life unattended? Scientists believe the library of life contains all gene and protein sequences, all metabolisms, and even all the regulatory signals required for life's unfolding. Did God commission another library to steer the unfolding of the species?

Neo-Darwinian evolutionists have failed to study the cosmos, ignoring cosmic blueprints and their otherworldly source. In just the past few years, modern DNA analysis is exposing their naivete. Unrelated species have been found to have identical DNA sequences—instructions for new species and complex adaptations—placing them in layaway in the quiet recesses of the genome.[191] These critical gene sequences gather in the quiet "nursery" of non-coding DNA, once considered "junk," until the time for their expression arrives. Life, we will soon see, draws on an eternal library of its own.

Let us finish this tour of the universe with take-home points to place the science of the universe in personal perspective:

- You were planned before the beginning of time.
- You are literally the product of light.
- To give you form and function, your light was converted into matter, quarks, atoms, elements, and then molecules that would someday create the miracle of life.
- Energy, including light, *cannot be destroyed*. It can morph into other forms of energy. It can be stored in batteries and fossil fuels and even in quarks. But light is immortal. What might that reveal about the reality of God and the nature of Heaven? We should listen to knowledgeable voices:

---

[191] K. Wang et al., "African Lungfish Genome Sheds Light on the Vertebrate Water-to-Land Transition," *Cell* 184, no. 5 (March 2021): 1362–1376.e1318, https://doi.org/10.1016/j.cell.2021.01.047.

From the Gospels

"I am the light of the world." John 8:12 (NIV).

"This is the message we have heard from him and declare to you: God is light; in him there is no darkness at all." 1 John 1:5 (NIV).

"You are the light of the world." Matthew 5:14 (NIV).

From the Apostle Paul

"About noon as I came near Damascus, a bright light from heaven flashed around me." Acts 22:6 (NIV).

"(Give) thanks to the Father, who has qualified you to share in the inheritance of the saints in the kingdom of light." Colossians 1:15 (NIV).

"For God, who said, 'Let light shine out of darkness,' made his light shine in our hearts to give us the light of the knowledge of God's glory displayed in the face of Christ." 2 Corinthians 4:6 (NIV).

"Wake up, sleeper, rise from the dead, and Christ will shine on you." Ephesians 5:14 (NIV).

From The Book of Revelation

"There will be no more night. They will not need the light of a lamp or the light of the sun, for the Lord God will give them light." Revelation 22:5 (NIV).

From People Who Had a Near-Death Experience, Describing a Brief Glimpse of Heaven

"And wonder of wonders, I realized I was seeing the inner light of all the growing things, just utter glory in color. It was not reflected light, but a gentle, inner glow that shone from each and every plant. Overhead, the sky was clear and blue, the light infinitely more beautiful than any light we know."[192]

"The landscape was beautiful, blue skies, rolling hills, flowers. All was full of light, as if lit from within itself and emitting light, not reflecting it."[193]

"I am instantly drawn toward the Light—I can feel its brightness, warmth, and love. As I get closer to it, I am absorbed by its brilliance and perfect love. Oh my God, I am the Light!"[194]

---

[192] Bruce Greyson, *After: A Doctor Explores What Near-Death Experiences Reveal about Life and Beyond* (New York: St. Martin Essentials, 2021), 34.
[193] Jeffery Long and Paul Perry, *Evidence of the Afterlife: The Science of Near-Death Experiences* (New York: HarperOne, 2011), 14.
[194] Long and Perry, *Evidence of the Afterlife*, 78.

"I look into the Light's source and see a massive, human silhouette that is radiating with the brightness of thousands of suns.... The Light knows me.... [There] are millions of other Lights welcoming me back home. I know them all and they know me; we are all pieces of the same Light."[195]

From light to life to eternal light, you are the product of God's sacred and eternal science.

The Book of God's Works and His Blueprints are about to open section 2, bringing the universe and our planet to life.

---

[195] Long and Perry, *Evidence of the Afterlife*, 79.

SECTION TWO
Life

Intro to Life

THE BLUEPRINTS (CODES) OF LIFE

These are the blueprints, or codes, commissioned to create life on Earth.

Complimentary Base-Pairing

There are four letters of DNA (four nucleotide bases: A, T, C, G). These four letters create a sentence three billion letters long in the genome of every cell. A (adenine) will only bind with T (thymine). C (cytosine) will only bind with G (guanine). The sequence A, T, C, G has the same biological meaning as T, A, G, C. As the double helix unwinds, complimentary base-pairing creates an identical copy of each strand as A's bind to new T's, T's bind to new A's, and C's and G's find new partners as well.

The Triplet Code

The triplet code supervises the construction of proteins. Any three of the four letters of DNA (ACG, TAC, etcetera) provide instructions to the protein factory (the ribosome) to add a specific amino acid to a growing protein chain. Proteins are long strands of amino acids joined head to tail, like snap beads, to construct the components of life. There are twenty amino acids (arginine, leucine, and tryptophan are examples). There are sixty-four three-letter "codons," more than enough to provide codons for all twenty amino acids.

Among their many activities, proteins build our tissues, transport raw materials around the cell, create nanomachines, receive and transmit signals, and function as antibodies to fight infection.

Each cell in our body contains more than forty million protein molecules.

Protein Folds

As a protein leaves the ribosome factory, it twists and vibrates into a unique three-dimensional structure that produces an "active site" that determines the protein's function. As an example, antibodies are proteins. Their active site binds to a target on an invading microorganism. Biologists are using artificial intelligence to understand protein folding and catalog active sites. Our body uses more than twenty thousand unique proteins, each with a unique fold and active site.

Regulatory Genes

Transcription is the process that copies gene sequences onto messenger RNA, sending instructions to the protein factory. "Transcription factors" controlled by regulatory gene complexes travel up and down the genome, turning protein-coding genes on or off.

Once considered "junk" because they don't code for a specific protein, regulatory genes control the tune and timing of the genomic orchestra.

Hox Genes

Hox genes are a subset of regulatory genes that play a special role in the journey of life. They divide the body into segments and ensure that the proper appendages are attached to each segment. There are *Hox* genes for eyes, antennas, wings, and limbs. They determine the type and shape of an embryo.

Given their command-and-control function in the development of an embryo, *Hox* genes might be considered life's version of cosmic blueprints. Regulatory genes control gene expression and the timing of protein production. *Hox* genes exhibit plug-and-play function and guide the creation of species.

A Sequence Library

Computer scientists have identified a library of gene and protein sequences stored in another dimension. That library organizes the biological hyperspace, a vast domain containing more potential proteins than atoms in the universe. The library also provides a search-and-retrieval system that allows a visiting gene to find a promising mutation.

While Darwinian "minor variation and natural selection" remains a valid explanation for incremental species modification (e.g., a lean pig or a woolly coat) and viral mutation (variants of COVID-19), a sequence library established before the beginning of time gives God command of the project, determining major directions.

Preassembly

Vast stretches of our DNA (98 percent, once considered "junk") provide a "nursery" for gene teams before the time for their expression arrives. A bilateral body plan, eyes, and claws for a Cambrian lobster, echolocation for bats and whales, and grazing teeth for the horse are a few examples of such expressions. The development of an extraordinary human brain is another example. Multi-genetic adaptations "preassemble" in the DNA nursery, then abruptly appear in the fossil record.

CHAPTER ELEVEN

Darwin and Beyond

Charles Darwin was no student of the cosmos. Except for Newton's gravitation and the orbital motion of planets, no biologist of his day had any concept of blueprints at work in the heavens. Darwin developed his theory of evolution through observation and travel, paying close attention to animal breeding and the diversity of species around the globe. As a result, he produced a powerful model of microevolution.

By building on existing species, minor variation and natural selection can create a stunning proliferation of life-forms and even generate new but closely related species.

But the most critical part of the story was missing. What produces new body plans and complex capabilities in the first place? Macroevolution, the development of novel body plans and complex adaptations, is the missing link in Darwin's theory.

Biologists Today Have No Excuse

Scientists have identified a team of blueprints interacting with symphonic precision to assemble a near-infinite cosmos. Those who would understand life should begin with the following premise: if there is a library of equations for the universe, there should be a corresponding library for life.

The significant events in life's journey, from developing bilateral body plans to seeing in the dark with sound to developing the mental ability to understand the math of black holes, demand a more fulsome explanation than chance-driven variation and selection. We should introduce two profound and recent observations.

The Discovery of a Library for Life

In a well-received book, *Arrival of the Fittest*, to be discussed in detail in chapter 14, computer scientists in Switzerland have described a library for life occupying a higher dimension.[196] While the mathematical landscape or library for the universe contains equations for every component of the cosmos, life's library contains equally critical information; it is a repository for all possible genes and proteins. "Neutral" ("non-coding") genes may visit its stacks and retrieve vital mutations for the future.

This Raises a Fascinating Question

How does life utilize information retrieved from that library?

Mutating genes may lay away the code for new genes and proteins in the quiet expanses of non-coding DNA. Less than 2 percent of our DNA is actively involved in the production of proteins for our cells and tissues. Teams of new genes may gather in the remaining 98 percent until the time for their expression arrives. Termed "preassembly" by Fredric Menger, professor of organic

---

[196] Andreas Wagner, *Arrival of the Fittest: Solving Evolution's Greatest Puzzle* (New York: Current, 2014), 14.

chemistry at Emory University, this process is presented in detail in chapter 14.[197]

This process would explain one of life's great mysteries. Why do most new species and complex adaptations show up fully formed in the fossil record? Trilobites appeared abruptly in the ancient oceans with no prior transitional fossils. Bats and whales suddenly developed the ability to hunt with sophisticated tools of echolocation. The genes for these creatures and capabilities slowly gathered in the non-coding genome until ready to make a fully formed appearance.

Sequencing of "Living Fossil" Genomes

Confirming the reality of this process and the critical role for preassembly, biologists have sequenced the largest genome of any known species, a "living fossil" called the lungfish, finding striking evidence for preassembly.[198] We will review their findings in detail in chapter 17.

---

[197] Fredric M. Menger, "An Alternative Molecular View of Evolution: How DNA was Altered over Geological Time," *Molecules* 25, no. 21 (November 2020): article 5081, https://doi.org/10.3390/molecules25215081.

[198] K. Wang et al., "African Lungfish Genome Sheds Light on the Vertebrate Water-to-Land Transition," *Cell* 184, no. 5 (March 2021): 1362–1376.e1318, https://doi.org/10.1016/j.cell.2021.01.047.

A New Theory of Macroevolution is Coming into Focus

Charles Darwin was a 19th century naturalist who wrote "Origin of the Species" in 1859. His theory of evolution focused on minor variation and the influence of natural selection.

We have seen that Darwin's theory applies well to the mutation of pandemic viruses as well as the breeding of animals and modification of existing species. His theory falls far short as an explanation for the abrupt appearance of new body plans and capabilities during the journey of life.

In his seminal work, *On the Origin of the Species*, Darwin asserted that life began in simple forms and evolved gradually over a prolonged period of time.[199] New species are produced by minor variations affecting the individual animal's ability to capture food,

---

[199] Charles Darwin, *The Origin of Species: 150th Anniversary Edition*, (New York: Signet, 2003).

survive, and generate offspring. Natural selection is the hero of the story.

If a variation promotes survival and reproduction, natural selection spreads that variation through the species. As variation accumulates, new species gradually emerge, making all creatures today descendants of ancient life-forms.

If farmers and naturalists can breed multiple generations of pigeons and pigs and shape their personality and appearance, nature can design every species. Darwin dismissed God from the process.

How have Darwin's original hypotheses withstood the test of time?

Darwin's Hypothesis Number One

Evolutionary change results from an accumulation of minor variations in a drawn-out, incremental process.

Darwin was so devoted to "incrementalism," he argued that if someone could lay out every species on Earth next to another, one generation followed by the next and then the next, it would be impossible to tell where one species ended and a new species began.[200] The fossil record, in other words, should overflow with transitional creatures.

Yet, *transitional fossils are rare.* Paleobiologists have done their best to identify transitional fossils. They suggest that Archaeopteryx is a dinosaur with feathers, a classic example of a creature in transition—a non-avian dinosaur on its way to becoming a bird.[201] A dog-like "Ambulocetus" walked along the shores before returning

---

[200]   John van Wyhe, "Darwin versus God?" *BBC History Magazine,* January 2009, 29, https://darwin-online.org.uk/content/frameset?viewtype=text&itemID= A669&pageseq=1.

[201]   Joseph Castro, "Archaeopteryx: The Transitional Fossil," Live Science, March 14, 2018, https://www.livescience.com/24745-archaeopteryx.html.

to the ocean as a whale[202] and to African rivers as a hippo.[203] But the fossil record, in general, has told a different story. New species appear abruptly with no partial mutants announcing their pending arrival.

The Cambrian Explosion

The unannounced appearance of new life-forms was particularly striking in the "Cambrian explosion," when multiple new body plans, complete with eyes and claws, appeared in the ancient oceans. Before the Cambrian, life consisted of primitive sponges and worms.[204] After the Cambrian, life had bilateral body plans and new appendages.

Biologists have spent decades digging through pre-Cambrian rock, searching for transitional precursors evolving into Cambrian species, with little to show for their effort.[205]

Harvard paleontologist Stephen Jay Gould, a strong advocate of Darwin's chance-driven incrementalism, lamented the weakness of the fossil record: "The absence of fossil evidence for intermediary stages between major transitions in organic design, indeed our inability, even in our imagination, to construct functional

---

[202] Nick Pyenson, *Spying on Whales: The Past, Present, and Future of Earth's Most Awesome Creatures* (New York: Viking, 2018), 28

[203] Fabrice Lihoreau et al., "Hippos Stem from the Longest Sequence of Terrestrial Cetartiodactyl Evolution in Africa," *Nature Communications* 6 (2015): article 6264, https://doi.org/10.1038/ncomms7264.

[204] Simon Conway Morris, *The Crucible of Creation: The Burgess Shale and the Rise of Animals* (Oxford: Oxford University Press, 1999).

[205] Allison C. Daley et al., "Early Fossil Record of Euarthropoda and the Cambrian Explosion," *Proceedings of the National Academy of Sciences* 115, no. 21 (May 2018): 5323–5331, https://doi.org/10.1073/pnas.1719962115.

intermediates in many cases, has been a persistent and nagging problem for gradualistic accounts of evolution."[206]

Unfortunately, Professor Gould died in 2002. What would he say today about the genes for life on land floating in the genome of ancient fish?

No Theory of the Generative

Thoughtful biologists have expressed their frustration with Darwin's model more clearly: "Sources of new form and structure must precede the action of natural selection."[207] Neo-Darwinism, in other words, lacks a compelling "theory of the generative." Some processes must nominate (generate) new genes and assemble them in teams to build a new species or a complex capability.

Where does life find the multiple moving parts for significant innovation? If Darwin's thesis is correct—that species and complex adaptations are the product of incremental minor variation one step at a time—why do they appear abruptly and fully formed in the fossil record?

Darwin's Hypothesis Number Two

Natural selection drives the creation of new species.

Evolutionists see natural selection as a "creative force," but natural selection is not creative.[208] The creation of new species and adaptations must happen in the genome. Should a mutation have

---

[206] Stephen Jay Gould, "Is a New and General Theory of Evolution Emerging?" *Paleobiology* 6, no. 1 (Winter 1980): 119–130, https://doi.org/10.1017/S0094837300012549.

[207] Stephen C. Meyer, "The Origin of Biological Information and the Higher Taxonomic Categories," Discovery Institute, August 4, 2004, https://www.discovery.org/a/2177/.

[208] Wagner, *Arrival of the Fittest*, 14.

a positive effect on an animal, natural selection can promote the distribution of that mutation. More often, it rejects genetic change.

Natural selection favors stasis, voting thumbs down on most mutations and producing prolonged periods of evolutionary inactivity. This natural process's resistance to change is one of the reasons that alligators have looked like alligators for eighty million years.

Natural selection is a powerful force for minor change, reshaping the texture of fur or the shape of a beak.[209] Variants of a pandemic virus demonstrate microevolution in action.[210] But the shuffling of a few letters of DNA hardly gives rise to new species.

Darwin's Hypothesis Number Three

New evolutionary traits will only appear in related species.

Darwin explained his theory with a drawing of a tree of life.[211] Arthropods and insects might occupy one branch of the tree and plants another, while birds and mammals sit on branches of their own. Darwin believed biologists could understand every species by its location on the limbs of his tree.

Modern gene sequencing has bulldozed Darwin's tree. The tree made sense when biologists determined the relationships among species by shared physical characteristics. But gene sequencing has demonstrated that physical features are an unreliable factor. Bats,

[209]   Jonathan Weiner, *The Beak of the Finch: A Story of Evolution in Our Time* (New York: Vintage Books, 1994).
[210]   Sovik Das, Swati Das, and M. M. Ghangrekar, "The COVID-19 Pandemic: Biological Evolution, Treatment Options and Consequences," *Innovative Infrastructure Solutions* 5, no. 76 (2020), https://doi.org/10.1007/s41062-020-00325-8.
[211]   Francis S. Collins, *The Language of God: A Scientist Presents Evidence for Belief* (Rockland, MA: Wheeler Publishing, Inc., 2007), 96.

whales, and mice could hardly be more physically different, yet they use the same genes to see in the dark with sound.[212]

Many biologists argue that the concept of a tree of life is obsolete.[213] As evolutionary scientist Eric Bapteste put it, "We have no evidence at all that the tree of life is a reality."[214]

Darwin's Hypothesis Number Four

Rewind the tape of life, and you will see an entirely different movie every time, according to Neo-Darwinian atheists.

Life has replayed its video many times, and the results are often the same. Life "converges."

Species find the same innovative ideas over and over. Simon Conway Morris, professor of evolutionary paleobiology at Cambridge University, praises the power of this "convergence," stating that even if the movie of life repeated itself over and over, "A lot of things will turn out the same way. Separate groups of organisms independently evolve similar solutions to similar problems, whether the solutions are teeth, eyes, brains, ecosystems, or societies."[215]

While biologists once attributed this convergence to the power of natural selection in response to environmental pressure, we will soon see that convergence often involves the utilization of similar, if not identical, genes.

---

212   Pyenson, *Spying on Whales.*
213   Ian Sample, "Evolution: Charles Darwin was Wrong about the Tree of Life," *The Guardian*, January 21, 2009, https://www.theguardian.com/science/2009/jan/21/charles-darwin-evolution-species-tree-life.
214   Sample, "Evolution."
215   Simon Conway Morris, *The Runes of Evolution: How the Universe became Self-Aware* (New Brunswick, NJ: Templeton Press, 2015), 21.

A Fair Challenge to Neo-Darwinism

It is time for Darwinists to answer two fundamental questions:

First, where do mutating species find the new genes, not just individual genes but gene combinations, required to take life in a new direction?

Second, where do new genes gather before the time for their coordinated expression? Isn't it likely there is a "nursery" in the genome where related genes can quietly gather until the ensemble is ready for action?

Thanks to the remarkable progress of "next-generation sequencing" (rapidly and efficiently identifying each letter in a genome), the answers to these questions are at hand.

The Human Genome Project

In 2000, US President Bill Clinton announced a remarkable scientific milestone: the Human Genome Project successfully read the first human genome—the complete set of genetic instructions for a human being.

The Human Genome Project was one of the most important feats of human exploration.[216] While cosmology and particle physics explored the universe, it was a voyage of discovery within, allowing humanity to sequence and read the book of life.

The project spawned fierce competition between government laboratories and private entrepreneurs, and this competition proved productive.[217] In the year 2000, one technician could sequence a

---

[216] Siddhartha Mukherjee, *The Gene: An Intimate History* (New York: Scribner, 2017).

[217] Viktor K. McElheny, *Drawing the Map of Life: Inside the Human Genome Project* (New York: Basic Books, 2012).

million DNA letters per day. By 2008, that same technician could sequence a billion letters per day.

Subsequent improvements in technology continue to advance the speed and accuracy of sequencing. Want to sequence your genome? Modern labs can accomplish the project for a few hundred dollars in a single day. Thirty years ago, it took ten years, multiple labs, and billions of dollars to sequence a single genome.

Why does the ability to rapidly determine the sequence of DNA matter? In just the past five years, scientists have been able to sequence the genome of many plant and animal species. The study of fossils has given way to the direct analysis of code. Next-generation sequencing has revolutionized our understanding of the species.

Sequencing and a genetic engineering device called CRISPR, a DNA slicing tool derived from bacteria protecting themselves from an invading virus, are rewriting the story of life. Stay tuned for some startling surprises.

Supplement Eleven

Examine how Darwin's changing views on life, design, and the origin of species affected his attitude toward faith.

CHAPTER TWELVE

DNA

Eight independent blueprints created a universe that defies our imagination. Then the unimaginable happened. Codes embedded in DNA made Earth a habitat for life.[218]

Three Codes in One Molecule

Thymine   Adenine

Guanine   Cytosine

Just four "letters" create multiple codes within the amazing molecule called DNA.

Key to its function is "complimentary base-pairing: Thymine will only bind with Adenine (T: A) and Guanine with Cytosine (G:C).

---

[218]   James D. Watson, *The Double Helix: A Personal Account of the Discovery of the Structure of DNA* (New York: Norton, 1968).

As hard as it is to imagine, the codes of DNA are written with just four letters, A, T, C, and G, representing four biomolecules known as nucleotides: adenine, thymine, cytosine, and guanine. Nucleotides of DNA store the code of life and obey a fundamental rule. A will only bind to T, and C will only bind to G. This complimentary arrangement creates two base-pairs that form the rungs of the double helix. As the two strands of the double helix unwind, A's reach for new T's, C's reach for new G's, and the original double helix becomes two. This process copies the three billion letters of our genome *two trillion times per day.*[219]

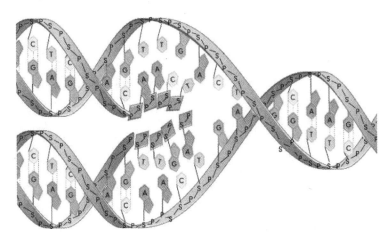

During DNA replication, the double helix unwinds. Adenine on each strand binds with a new thymine (and vice versa), while cytosine binds with a new guanine. The original helix visible on the far right becomes two new helices to the left.

---

[219] Sarah Zhang, "Your Body Acquires Trillions of New Mutations Every Day: And It's Somehow Fine?" *The Atlantic*, May 7, 2018, www.theatlantic.com/science/ archive/2018/05/your-body-acquires-trillions-of-new-mutations-every-day/559472/.

Each cell in our body contains a run-on sentence of DNA three billion letters long. Even more remarkably, that run-on sentence of nucleotides or "bases" delivers three separate ingenious codes.

As briefly outlined above, one code duplicates DNA, a critical step in delivering a genome to a daughter cell. A second code, described below, supervises the construction of the strings of amino acids called proteins that provide the tissues and functions of life. A third code, also summarized below, regulates the on and off actions of genes, a vital code as a four-cell embryo becomes a fully formed human infant. *Hox* genes are commanders of the regulatory process, segmenting the body plan and attaching appendages, including limbs, antennas, and eyes.

The four letters of DNA divide into sixty-four three-letter codons and supervise the construction of proteins. Proteins and their active sites provide the structure and function of life.[220] As proteins (chains of amino acids attached nose to tail, like snap beads), exit the ribosome, they twist and fold into a final shape.

Finally, master control genes (*Hox* genes, a powerful subset of regulatory genes) dispatch their "transcription factors" up and down the genome, placing certain genes in "transcription" mode (copies the three-letter codon sequences onto messenger RNA and send the instructions to the protein factory). Alternatively, regulatory factors might deliver an entirely different message: it's time to coil up your double helix and go silent.[221]

In a spasm of elegant activity, regulatory genes convert an amorphous embryo into a fully formed infant. There are regulatory genes for every organ and appendage, creating the cell types

---

[220] Watson, *Double Helix.*

[221] Pamela J. Mitchell and Robert Tjian, "Transcriptional Regulation in Mammalian Cells by Sequence-Specific DNA Binding Proteins," *Science* 245 no. 4916 (July 1989): 371–8, https://doi.org/10.1126/science.2667136.

required for each component of life, from organs and eyes to fingers and toes.[222]

Once labeled "junk," *regulatory genes may be the most important DNA code of all.* See chapter 13.

> **Question:**
> Is it simply good fortune that DNA carries three independent and crucial codes in one molecule capable of duplicating DNA, supervising protein production, and regulating gene expression?

The Miracle of Life

Many of us behold a newborn infant as a gift from heaven. Imagine the miraculous activity that converts a four-cell embryo into a fully formed human being. Master genes dispatch organ-specific stem cells to their target tissue: build a heart here, limbs there, and begin forming eyes and ears and a brain. Within each organ, DNA dispatches an armada of messages to make new cells at a furious pace, assuring that each cell contributes to its team. It gives each cell its own book of life, all three billion letters of DNA.

Each cell then build tens of thousands of proteins. Proteins are our architects, building our tissues, muscle, and hair. Proteins are our chemists, speeding up chemical reactions. They create the molecular nanomachines that drive the activities of the cell. They also import and export molecules across the cell membrane and straddle the membrane, listening for messages from other cells.

---

[222] Jean S. Deutsch, ed., *Hox Genes: Studies from the 20th to the 21st Century*, Advances in Experimental Medicine and Biology, vol. 689 (New York: Springer, 2010), 689.

If the process was accompanied by music, expectant mothers would be serenaded by a non-stop symphony unmatched by the world's best philharmonic.

Key Definitions:

Complimentary Base-Pairs
Adenine will only bond to thymine. Cytosine will only bind to guanine. Two base-pairs, A-T and C-G, carry the code of life.

Transcription
The process by which the codons for a specific protein are copied from the DNA sequence onto letters of messenger RNA.

Translation
Protein factories called ribosomes read the codons of messenger RNA like ticker tape, building proteins one amino acid at a time.

The Magic of Protein Folding

The protein factory (ribosomes) read codons one amino acid at a time. But when the protein exits the ribosome, it undergoes a dramatic transformation, collapsing like spaghetti into a wondrous world of origami. By folding into a precise three-dimensional shape, the protein creates an "active site" that executes the protein's function.[223] The active sites of antibodies latch onto an invading microorganism. Active sites formed by the proteins of muscle (myosin and actin) behave like ratchets and move our limbs. The

---

[223] Eric T. Kool, "Active Site Tightness and Substrate Fit in DNA Replication," *Annual Review of Biochemistry* 71 (July 2002): 191–219, https://doi.org/10.1146/annurev.biochem.71.110601.135453.

active sites of other proteins ferry raw materials across the cell, slice molecules into smaller parts, or speed up chemical reactions.

Nanomachines

In a process that strains our imagination, teams of proteins form nanomachines in our cells that resemble the most sophisticated products of human engineers.[224] As explained in detail in supplement 15, the flow of protons through a protein called ATPase turns a rotor 130 revolutions per minute like a man-made turbine, storing the chemical energy for the ongoing demands of life.

Each cell contains up to ten million ribosomes and over forty million protein molecules jostling with each other like bumper cars.[225] It is no surprise that scientists have described a cell's proteins as commuters in a Tokyo subway at rush hour.[226] Ribosomes are nonstop beehives of activity, continuously printing out new proteins.

What are the Odds?

How many proteins can three hundred amino acids build, with twenty different amino acids available for each slot on the chain? The answer is staggering: twenty raised to the three hundredth power (20^{300}). If just two amino acids could build a protein with

[224] Mohinder Pal et al., "Structure of the TELO2-TTI1-TTI2 Complex and its Function in TOR Recruitment to the R2TP Chaperone," *Cell Reports* 36, no. 1 (July 2021): article 109317, https://doi.org/10.1016/j.celrep.2021.109317.

[225] Brandon Ho, Anastasia Baryshnikova, and Grant W. Brown, "Unification of Protein Abundance Datasets Yields a Quantitative *Saccharomyces cerevisiae* Proteome," *Cell Systems* 6, no. 2 (February 2018):192–205.E3, https://doi.org/10.1016/j.cels.2017.12.004.

[226] Andreas Wagner, *Arrival of the Fittest: Solving Evolution's Greatest Puzzle* (New York: Current, 2014), 63.

twenty amino acids available for each slot, there would be four hundred proteins (20^2). Add a third amino acid with twenty more options, and we are up to eight thousand (20^3). So, it is easy to see that when you build a protein three hundred amino acids long, the number of possibilities is 20^{300}. This vast world of proteins, known in scientific terms as the proteome, represents the hyperspace of biology.[227] The number of potential proteins in the universe far outnumbers the number of atoms (a mere 10^{80}).

It challenges common sense to believe that a cell finds the perfect protein by randomly wandering through the protein space. And yet, Darwinian theory demands that proposition. In the Darwinian view, life navigates the hyperspace of proteins while wearing blindfolds as natural selection guides the ship.[228]

There must be a more elegant explanation, and leading scientists have detected its outline. In a stunning book, *Arrival of the Fittest*, computer-armed biologists have identified a library of sequences for life. That library occupies a higher dimension, precisely like the mathematical landscape (library of blueprints) we met in section 1.[229] We will visit life's library in chapter 14. Its structure will challenge your imagination.

Hints of the Same Designer

In blueprints of the universe, quarks came in two flavors, up and down. DNA codes involve two flavors of base-pairs, A-T and C-G. Both the code of the quarks and the base-pairs of DNA involve two units (up and down, A-T and C-G).

---

[227] Wagner, *Arrival of the Fittest.*
[228] Charles Darwin, *The Origin of Species: 150th Anniversary Edition*, (New York: Signet, 2003).
[229] Wagner, *Arrival of the Fittest.*

In codes of the universe, the combination of three quarks builds protons and neutrons for the atom. In the codes for life, three letters (a "triplet code") build codons that determine the sequence of amino acids in proteins.

Unsolved Mysteries: The Formation of DNA and the Origin of Life

Scientists have tried to prove that DNA self-assembled somewhere on early Earth. Not only have they failed to find evidence for the spontaneous self-assembly of DNA, but they have also failed to find evidence for the self-assembly of any single letter of DNA.[230] Biologists have only been able to find DNA and its four nucleotide letters within living cells.

No nonliving system on a clay surface, at a deep ocean vent, or in a warm, little pond has produced these molecules of life. How did life manage to put them together?

We should acknowledge another, equally confounding mystery. How did life arise, complete with the ability to replicate its code and manage its energy budget? If it is difficult to explain the origin of DNA, how can we produce a plausible explanation for the origin of life?

The codiscoverer of DNA, Francis Crick, was so frustrated by our inability to explain the origin of DNA that he suggested the molecule arrived on a rock from outer space.[231] Seeing no way that DNA could have spontaneously formed on our planet, he proposed that DNA assembled somewhere else in the cosmos and hitchhiked

---

[230]   Orson Wedgwood, *DNA: The Elephant in the Lab—The Truth about the Origin of Life* (self-pub., Wedgwood Publishing, 2019), 52.
[231]   Francis Crick, *Life Itself: Its Origin and Nature* (New York: Simon and Schuster, 1981).

its way to our planet. This hypothesis—called panspermia—pushed the challenge out of this world.

The universe had a beginning in an inexplicable eruption of light. Science lacks the tools to decipher the moment of creation or understand the source of its contents.[232]

Like the universe, life and DNA had a beginning in our planet's deep history. And once again, those beginnings defy the best of science. Decades ago, scientists believed they would explain the origin of life. Today, those scientists remain dumbfounded.[233]

Scientists worldwide recognize that the statistical and conceptual hurdles are so immense that it is unlikely we will ever develop a plausible explanation for the origin of life. The blurb of *The Fifth Miracle*, a book that analyzed the latest life origin theories at the time, summarized the status of the field: "The origin of life remains one of the great unsolved mysteries of science."[234]

God of the Gaps

Physicists readily admit that they do not have the tools to explore a point of infinite heat and density that describes the big bang.

Despite the persisting mystery of the origin of life, some scientists continue to claim an explanation may be close at hand. Give us more time, some say; we may find an answer tomorrow. They warn by attributing the creation of life and the synthesis of DNA to the

---

[232] Roger Penrose, *The Road to Reality: A Complete Guide to the Laws of the Universe* (New York: Vintage Books, 2004).
Andreas Wagner, *Arrival of the Fittest: Solving Evolution's Greatest Puzzle* (New York: Current, 2014), 691.
[233] Change Laura Tan and Rob Stadler, *The Stairway to Life: An Origin-of-Life Reality Check* (self-pub, Evorevo Books, 2020).
[234] Paul Davies, *The 5th Miracle: The Search for the Origin and the Meaning of Life* (London: Penguin, 2003).

creative power of God, we may be attributing a natural phenomenon to a "god of the gaps."[235]

This book does not see a god in the gaps. It sees Him in the brilliance of His sacred science.

We Have Met the First Three Blueprints (Codes) for Life

The code of complimentary base-pairing
The triplet code
The code of protein folds

Supplement Twelve

In the 1960s, three separate groups competed to be the first to decipher the structure of DNA. Review the dramatic details of this competition as scientists attempted to model the double helix and identify the molecules that carry the code of life.

---

[235] Francis S. Collins, *The Language of God: A Scientist Presents Evidence for Belief* (Rockland, MA: Wheeler Publishing, Inc., 2007), 93.

CHAPTER THIRTEEN

Regulation

We have met the blueprints for DNA duplication and protein production and have reviewed the importance of protein folds. In this chapter, we meet the master genes supervising the process. Master control genes regulate the activity of protein-coding genes, turning them on or off.

Gene Regulation

We have previously imagined the beautiful symphony that might accompany the development of a human fetus. Trumpets announce the marriage of two gametes (male and female germ cells). In minutes, a brand-new embryo becomes four cells, then quickly eight. Violins celebrate the layout of a body plan, while trombones welcome the appearance of organs.

The heart needs a particular repertoire of proteins, while lungs need a separate set entirely. Orders fire up and down the chromosome as messenger RNA flows out of the genome like notes from a concert piano.

Hox Genes

Master regulatory genes (*Hox* genes in the terminology of embryology) are the conductors of the DNA orchestra.[236] By dispatching "transcription factors" to turn genes on and off, these master control genes supervise the body plan and the construction of organs and appendages. The human eye, for instance, has a dedicated control gene called *PAX6*.

Parents behold a newborn as a miracle for a good reason. The regulatory code that supervised that infant's construction was executing God's intentions.

Wolves on Continents Apart

The Australian wolf and the North American wolf evolved on separate continents with no common ancestor in one hundred and sixty million years. And yet, these wolves have an identical shape to their skulls.[237] Neo-Darwinists proposed that natural selection determined the formation of both skulls. Survival of the fittest, the story goes, demanded identical jaws to catch and eat rabbits and wombats.

Modern DNA analysis has demonstrated the truth of the matter.[238] The shape of these skulls reflects the activity of near-identical

---

[236] M. Mark et al., "*Hox* Gene Function and the Development of the Head," in *Neural Cell Specification*, ed. Bernhard H. Juurlink et al., (Boston: Springer, 1995), 3–16, https://doi.org/10.1007/978-1-4615-1929-4_1.

[237] M. R. G. Attard et al., "Skull Mechanics and Implications for Feeding Behavior in a Large Marsupial Carnivore Guild: The Thylacine, Tasmanian Devil and Spotted-Tail Quoll," *Journal of Zoology* 285, no. 4 (August 2011): 292–300, https://doi.org/10.1111/j.1469-7998.2011.00844.x.

[238] Charles Y. Feigin, Axel H. Newton, and Andrew J. Pask, "Widespread *cis*-Regulatory Convergence between the Extinct Tasmanian Tiger and Gray Wolf," *Genome Research* 29 (2019): 1648–1658, https://doi.org/10.1101/gr.244251.118

regulatory genes. Natural selection has no power to nominate or produce a gene.

> **Question:**
> *How did these unrelated creatures acquire identical genes more than two hundred million years in the past to direct the shape of their skull?*

Surprises from the First Human Genome

The Human Genome Project shocked the world with its initial findings. Three billion letters long, our genome could potentially produce more than three hundred thousand genes, containing nine hundred nucleotide letters per gene. But sequencing found a much smaller number of "coding genes." Only twenty to twenty five thousand genes, less than 2 percent of our genome, provide the "coding DNA" for our tissues and cells.[239] We share many of these coding genes with numerous animal species, including mice and chimps.

For decades, scientists have mislabeled the remaining 98 percent of our DNA as "junk." This label is a leading candidate for biology's number one misnomer. "Non-coding DNA" is not junk. To the contrary, it represents one of life's treasures: it houses regulatory genes. Regulatory genes don't send messages to the ribosome or produce individual proteins for cells and tissues. Instead, they exert their function within the genome, controlling the timing of gene expression. Think of control genes as the commanding officers of

---

[239] Francis S. Collins, *The Language of God: A Scientist Presents Evidence for Belief* (Rockland, MA: Wheeler Publishing, Inc., 2007), 124.

the genome, telling a gene to unwrap its sequence for transcription or, conversely, to coil up and stop production.[240]

> **Key Definition: Transcription Factor**
> *A protein or RNA sequence that binds to a gene to open the helix and transcribe the nucleotide sequence onto a single strand of messenger RNA. Transcription factors can also tell the gene to coil up and turn off.*

Regulatory genes oversee the layout of body plans, the construction of organs and eyes, and the placement of antennas and wings. If "coding" DNA provides the nuts and bolts for body construction—the instruments of the development orchestra—*Hox* genes are the orchestra's conductor.[241]

Regulatory codes turn coding genes on or off in tissues of the developing fetus. Place the heart here. Put a wing over there. Start the construction of the jaw and the formation of teeth here. Regulatory code provides the development plan for every fetus.

Just twenty years ago, neo-evolutionists attributed the development of the eye to hundreds of small steps supervised by natural selection.[242] As a first step, according to their just-so story, a simple light-sensitive spot on the skin of some ancestral creature allowed it to escape a predator. Then, over time, a pit formed in the light-sensitive patch, making it possible to respond more quickly

---

[240] Jean S. Deutsch, ed., *Hox Genes: Studies from the 20th to the 21st Century,* Advances in Experimental Medicine and Biology, vol. 689 (New York: Springer, 2010), 689.

[241] Deutsch, *Hox Genes.*

[242] Richard Dawkins, *The Blind Watchmaker: Why the Evidence of Evolution Reveals a Universe Without Design* (New York: Norton, 1996).

to the light. The pit's opening gradually narrowed, creating a small aperture—evolution's version of a pinhole camera.

No matter how slight, every change improved the odds of the creature's survival. Eventually, the light-sensitive spot evolved into a retina, while a lens formed at the front of the eye. Hundreds of different steps, guided by natural selection, produced the eye, not only once, the story goes, but over forty times in unrelated creatures.

It is difficult not to laugh at this just-so story. While the step-by-step process may have happened once in the ancient ocean, master control genes such as *eyeless* and *PAX6* regulate the construction of eyes in all forty species.[243] These are boss genes with overriding powers. They can turn on all the genes necessary to build a working eye. The same code has overseen the construction of eyes in insects, the octopus, and a wide variety of mammals including humans. *If these master regulator genes for the eye express themselves in the wrong tissue, they can convert legs and wings into eyes.*[244]

> ### Key Distinction:
> *The terms "coding" and "non-coding" refer to the process of protein production. Only 2 percent of our genes code for the production of proteins.*
> *Ninety-eight percent of our DNA does not code for protein production. Once considered "junk," regulatory genes live amid "non-coding" DNA. Other stretches of "neutral" or "non-coding" DNA may provide the "nursery" for the gathering of future genes.*

---

[243] J. L. Zagozewski, Q. Zhang, and D. D. Eisenstat, "Genetic Regulation of Vertebrate Eye Development," *Clinical Genetics* 86, no. 5 (August 2014): 453–460, https://doi.org/10.1111/cge.12493.
[244] Mark et al., "*Hox* Gene Function."

The discovery of *Hox* genes by developmental biologists in the 1980s provided DNA research with its "Rosetta stone." A "*Hox* toolbox" filled with highly conserved regulatory genes steers the ship of a species. Instead of minor variation as the secret to life, master control genes provide a more powerful explanation. The journey of life involves a toolbox. What craftsman designed the tools?

The Significance of Hox Genes

Multilayered *Hox* gene complexes preserved across eons of evolutionary time present a significant challenge to neo-Darwinian theory. In the traditional Darwinian model, the emergence of a new species involves minor variations supervised by natural selection. Innumerable variations guided by selection slowly separate a new species from its ancestral pool.

The real story is vastly different. Microevolutionary changes within a species is one thing, macro is another. Life stores its best ideas in regulatory code. Build the eye this way, and shape the skull that way. Unrelated animals like Australian and North American wolves find near-identical regulatory genes defining the shape of their skull. Life leaps from branch to branch of Darwin's discredited tree.

The Story of Macroevolution

Where are we now with the ingenious blueprints of life? One blueprint supervises the replication of DNA, and one double helix becomes two. A replicated genome provides the DNA for a daughter cell or transmits a complete set of codes to the next generation.

A second blueprint determines the sequence of a protein. Attach these amino acids end to end, have them fold this way, and create an enzyme or an antibody and build the tissues of life.

Now, a third DNA code, the code of regulation, builds watch-towers in the genome. This code controls the fetus's layout, the body's segmentation, and the design of organs and eyes.

A central question remains. Through what scientific process does God influence the origin of the species and gently influence life's direction? Answers to this vital question are coming into focus.

As we will review in detail in the next two chapters, God placed potentially useful gene sequences, proteins, and regulatory codes in a library of potential mutations. Mutating genes visit that library to obtain their future options (chapter 14).[245] Sequences for a new function or a new species quietly organize themselves in the stretches of "junk" DNA in a process best described as "preassembly" (chapter 15).[246]

We Have Met an Additional Blueprint for Life

The code of complementary base-pairing
The triplet code
The code of protein folds
Regulatory genes

---

[245] Andreas Wagner, *Arrival of the Fittest: Solving Evolution's Greatest Puzzle* (New York: Current, 2014).

[246] Fredric M. Menger, "An Alternative Molecular View of Evolution: How DNA was Altered over Geological Time," *Molecules* 25, no.21 (November 2020): article 5081, https://doi.org/10.3390/molecules25215081.

Supplement Thirteen

After the ancient continent Pangea broke into separate land masses, Australian and North American species have developed independently for another two hundred million years. And yet, species on both continents can look like kissing cousins. Flying squirrels from both continents can be difficult to tell apart, even though the Australian varieties (marsupials) raise their young in a pouch while their North American counterparts nurture fetal development in a placenta.

The Australian wolf called a thylacine (extinct since the 1930s) and the North American wolf have an identical shape to their skulls. Evolutionists long attributed their similar skulls to the power of natural selection. Modern sequencing has revealed the truth of the matter: both species have identical regulatory genes determining the shape of their skulls.

CHAPTER FOURTEEN

Library for Life

Rather than make life wander through the biological hyperspace inspecting trillions of proteins, a sequence library provides a search system for the acquisition of useful genes. Mathematicians explore an eternal landscape to discover blueprints of the universe. Neutral, non-coding genes make a similar journey, retrieving new sequences for life.

The Discovery of a Library for Life

A library of mathematical blueprints defined the universe before the beginning of time. Many mathematicians see themselves as explorers, visiting that mathematical landscape through their mind's eye and retrieving insight to decipher the cosmos.[247]

Would God create a library for His universe, then ignore the needs of life? Not a chance. Mutations define life's future, its adaptations, and the development of new body plans and complex nervous systems. A library for life provides a road map for the species and facilitates the retrieval of mutations.

---

[247] Patchen Barss, "How a Silence Solved the Weird Maths Inside Black Holes," BBC, October 9, 2020, www.bbc.com/future/article/20201008-the-weird-mathematics-that-explains-black-holes-exist.

Supercomputers Have Outlined Life's Library

Supercomputers have revealed the outline of life's library, identifying its vast collections of metabolisms, gene networks, proteins, and regulatory circuits.[248]

How might God organize the sequences for life? Imagine a gift box with a distinctive feature: rather than limiting itself to function as a single container, it contains a second box, a cube within a cube, a geometric structure known as a "hypercube." A square has four corners, a cube eight, while a hypercube has sixteen.

Imagine the library as fifty cubes within cubes, a geometric structure fifty hypercubes tall, each cube containing another. Such a structure has more corners than we can imagine: a hyper-astronomical number of 25000 corners!.[249]

Who visits such a library in real life? Mutating genes travel its hypercube highways (possibly through the power of quantum superposition), bringing innovative ideas home to their genome. Unrelated wolves utilize identical regulatory genes to determine the shape of their jaws. Mammalian nervous systems grow increasingly complex until one privileged species connects with God's blueprints and purpose.

Question:
Hypercubes! A library for life in a higher dimension! Is this chapter a flight of fancy? Have supercomputers really identified life's library?

---

[248] Andreas Wagner, *Arrival of the Fittest: Solving Evolution's Greatest Puzzle* (New York: Current, 2014).
[249] Wagner, *Arrival of the Fittest*, 90.

The Library for Life

Using powerful computers and massive data sets, scientists have described a library of life's sequences (gene networks, regulatory genes, and proteins) in a fifty-dimensional hypercube established in a higher dimension. See Andreas Wagner's 2015 book, *Arrival of the Fittest: How Nature Innovates* for more information.

Before and After the Cambrian Explosion

Just four letters of DNA and twenty flavors of amino acids have produced untold millions of species. Those species utilize vast numbers of metabolisms and untold numbers of molecular machines. How does life navigate the enormous biological cosmos, populated with more potential entries than there are atoms in the universe? And how might God influence the navigation?

Neo-Darwinists have told us many times that evolution is an accidental process. Genes mutate. Mutation drives a random process involving tiny sequential steps. Natural selection votes on the value of each modification by determining the viability of offspring.

Microevolution of Anole Lizards

That formula works well in the domain of microevolution. A single species of anole lizard transplanted to a new Caribbean island quickly fills every environmental niche. In just one to two decades, a breeding pair of lizards gives way to new species living on the ground, occupying tree limbs, and even climbing high in the trees, each with minor changes to their claws and legs.[250] In these lizards, a repertoire of regulatory genes was preassembled and ready for action.

Darwinism successfully explains this process of micromutation, such as changes in the length of a bird's beak, the shape of a lizard's leg, or the impact of breeding on pigeons, pigs, and other domestic species. Minor change and natural selection can also produce new species if a small number of mutations change an important structure.

But a piecemeal process of random variation seems inadequate to explain the development of a complex new function. How could random mutations assemble the numerous parts required for a multigenic adaptation, such as the ability to see in the dark with sonar (echolocation)? As we will see in a future chapter, echolocation requires modifications to the organs that produce sound, the organs that acquire the returning echoes, and the brain that interprets the echoes. Despite this complexity, a bat can fly inches above your head in total darkness and barely disturb your hair. What are the odds that random mutations would produce this amazing adaptation, much less an entirely new body plan or species? God created a way to accelerate the discovery of useful mutations and channel the flow of life.

Cambrian Bust

Darwin's theory faced a challenge as soon as he published *On the Origin of Species*. It failed to explain the wondrous proliferation of

---

[250] Terry J. Ord, Judy A. Stamps, and Jonathan B. Losos, "Convergent Evolution in the Territorial Communication of a Classic Adaptive Radiation: Caribbean *Anolis* Lizards," *Animal Behavior* 85 no. 6 (June 2013): 1415–1426.

entirely new species in the ancient seas, referred to as the "Cambrian explosion,"[251] a shortfall Darwin himself acknowledged.[252]

Before the Cambrian era, life consisted of simple bacteria, blob-like jellyfish, and sponges breathing through their skin on the ancient ocean floor. In a whisker of evolutionary time, entirely new species erupted on the scene with bilateral body plans and striking appendages, including eyes, flippers, and claws.

Biologists have searched through multiple layers of rock produced by those ancient oceans, identifying fossilized shrimp, lobsters, insects, and many other new creatures. Almost all body plans used by life today made a debut in the Cambrian seas. One animal called the trilobite resembles modern-day woodlice or "roly-polies," except they often weighed more than ten pounds.[253]

Did precursor fossils provide advance notice that these splendid species were under construction? Not a hint. Despite more than a century of searching, biologists have failed to find transitional fossils that preceded the Cambrian creatures.[254] This failure to explain the abrupt appearance of new life-forms in the Cambrian era invalidated Darwinism at its inception, and even Darwin acknowledged the problem.[255] It is one thing to reshape the leg of a lizard or the length of a beak. It is something else entirely to build a new body plan adorned with eyes, antennas, and claws.

---

[251] Simon Conway Morris, *The Crucible of Creation: The Burgess Shale and the Rise of Animals* (Oxford: Oxford University Press, 1999), 141.

[252] Riley Black, September 2009, *Did the Cambrian Really Give Darwin Nightmares?* https://www.nationalgeographic.com/science/article/did-the-cambrian-really-give-darwin-nightmares

[253] Stephen Jay Gould, *Wonderful Life: The Burgess Shale and the Nature of History* (New York: W. W. Norton & Company, 1990), 16.

[254] Hou Xian-Guang et al., *The Cambrian Fossils of Chengjiang, China: The Flowering of Early Animal Life*, 2nd ed., (Hoboken, NJ: Wiley, 2017).

[255] Charles Darwin, *The Origin of Species: 150th Anniversary Edition*, (New York: Signet, 2003), 340.

A New Model

The time has come to embrace the new theory of life's unfolding; yet the concept of a mutational library is not for the faint of heart.

In prior chapters of this book, we met one of life's most significant debates. Do human beings invent mathematical equations? Or do they retrieve those equations from a mathematical dimension? Recall that a mathematical equation led to the discovery of the Higgs field, the encoding cosmic ocean that gives the universe physical structure. Did Peter Higgs invent the equation, or did he chance upon it and recognize its significance with the God-given capacity of his mind?

Higgs hardly invented the equation. It is older than the universe itself. Leading mathematicians like Sir Roger Penrose believe the Higgs equation and other mathematical formulas populate an invisible landscape of ideal forms.[256] In this reference, Penrose describes the power of Plato's library as follows: "Plato's world is an ideal world of perfect forms, distinct from the physical world, but in terms of which the physical world must be understood... It is this potential for the 'awareness' of mathematical concepts involved in this Platonic access that gives the mind a power beyond what can ever be achieved by a device dependent solely on computation for its action. Mutating genes might visit life's library every day.

---

[256] Roger Penrose, *Shadows of the Mind: A Search for the Missing Science of Consciousness* (Oxford: Oxford University Press, 1994), 50.

> **Key Definition: Mutation**
> *A change in the amino acid sequence of a gene caused by (1) the alteration of a single nucleotide or (2) by the deletion, insertion, or rearrangement of larger sections of genes or chromosomes. New research suggests another mechanism: a neutral, non-coding gene can mutate into an alternative sequence, providing a vital source for new genetic direction. A library for life might provide the source.*

Mathematicians hardly live or die based on their ability to visit another dimension. Species exploring the mutational library might put their life on the line with every visit. Should a coding gene required for the needs of the cell incorporate a harmful mutation, the species could face extinction. Make one wrong turn with a coding gene, mutate into a cul-de-sac, and a mutation could threaten your species.

But mutating genes in quiet, non-coding DNA ("neutral mutations") could make repeated visits with no risk to the species' future. Even more, the mutational landscape harbors a secret. God provided a system that increases the value of a visit. A network of superhighways crisscrosses the hypercube library, exposing genes to mutational options.[257]

Traveling the Mutational Landscape

The gene, of course, has no idea it is traveling a mutational landscape. Metaphorically, it parachutes onto the landscape, mutates its code, and returns to the safety of its genome. But an advance party

---

[257] Wagner, *Arrival of the Fittest.*

has maximized the value of a visit. Most multicellular creatures, for instance, including birds and mammals, base their metabolism on glucose. These species can freely mutate along a glucose superhighway, one harmless mutation at a time, visiting substantial portions of the sequence space and capturing helpful mutations while they travel.[258]

Networks provide safe passage through the landscape, but they do even more. Networks intersect with each other. The "metabolize glucose" network might intersect with the network of "downy feathers." Not only has a species modified its metabolism in this hypothetical instance, but it has bulked up its cold weather protection. Networks intersect at multiple locations, making beneficial code just a few mutational steps away.[259]

Supercomputer Sight

How do we know that these mutational landscapes exist? Biochemists have mapped thousands of metabolisms, outlining unique ways to synthesize macromolecules and obtain energy. Molecular biologists have deciphered untold numbers of gene and protein sequences. Other scientists have built massive data sets focused on regulatory gene circuits.

Then, a leading team of computer scientists decided to examine this data in a new and powerful way. They combined all these data sets and explored them with a new set of eyes: an array of supercomputers. It was time to see the forest through the trees.[260] Their findings? As mentioned earlier, a sequence landscape with

---

258 Wagner, *Arrival of the Fittest.*
259 Wagner, *Arrival of the Fittest.*
260 Wagner, *Arrival of the Fittest.*

potential mutations woven into a fabric more complex than human beings can imagine.[261]

An Endless Landscape of Opportunity

Neo-Darwinism would suggest that a species finds a new regulatory complex or a new protein through a random search. Now science is beginning to understand the plan: an unimaginable mind rigged the search process. Helpful highways and intersections are scattered across the landscape, providing multiple "needles in the haystack."[262]

This library or landscape of life is invisible to the human eye. But computer technology has provided a partial map of its structure, revealing, in the words of Wagner, "a Platonic world of crystalline splendor, the foundation of life's innovability."[263] Networks of genes and metabolisms, proteins, and regulatory elements "exist in the timeless, eternal realm of nature's libraries."[264] Neutral mutations browse the library, safely returning to their genome and accelerating innovation "like the warp drives of Star Trek."[265]

Potential Role of Quantum Superposition

Some scientists wonder if quantum mechanics plays a role in this process. Do gene sequences enter a superposition and entangle with multiple potential mutations during their library visits?[266] The quantum nature of fundamental particles was essential to God's

---

[261] Wagner, *Arrival of the Fittest*.
[262] Wagner, *Arrival of the Fittest*.
[263] Wagner, *Arrival of the Fittest*.
[264] Wagner, *Arrival of the Fittest*.
[265] Wagner, *Arrival of the Fittest*.
[266] Johnjoe McFadden, *Quantum Evolution: The New Science of Life* (New York: W. W. Norton and Company, 2001).

work in the cosmos. Through superposition, the wave function of a particle gives it the freedom to occupy unlimited positions in the same instant. Entanglement allows particles to join forces. These quantum phenomena play vital roles during nuclear fusion and the creation of higher elements.

Beryllium and carbon, for instance, make the same quantum leap in the stars. They briefly entangle with each other until carbon bypasses beryllium and maximizes its production. Without quantum superposition and entanglement, Earth might still be waiting for life.

Perhaps mutating genes enter a quantum superposition of options—a "quantum forest of possible gene sequences."[267] The mutation emerging from this entanglement could deliver a new and improved capability to the genome.

Just like the Higgs equation, no anatomist can access these landscapes with their hands. No microbiologist can see them with their microscopes. It took the power of the world's fastest computers to identify their outline. As Wagner describes them, "They are concepts, mathematical concepts, touchable only by the mind's eye."[268]

What mind could have envisioned them in their totality? The coding landscape, like its mathematical counterpart, bristles with divine implications.

We might refer to this hypercube collection of gene sequences by giving it a familiar name:

Library for the Origin of Species.

---

[267] McFadden, *Quantum Evolution*, 269.
[268] Wagner, *Arrival of the Fittest*, 291.

We Have Met an Additional Blueprint for Life

The code of complementary base-pairing
The triplet code
The code of protein folds
Regulatory genes
A library of gene and protein sequences

Supplement Fourteen

Many human beings would envision a mathematical library as a building filled with shelves of books containing hard-to-understand equations and theorems. But the sequences of DNA and proteins are mathematical as well. They exist not only in landscapes in a higher dimension but physically in the cells and tissues of living creatures.

Great cultures have long been fascinated by the power of libraries. Review the great libraries of the ancient past. Few man-made libraries survive from ancient times. In contrast, the divine libraries of math and life are eternally stable.

CHAPTER FIFTEEN

Preassembly

The discovery of a library for life housed in another dimension raises fundamental questions. How does life retrieve library information? How does it put the information to work?

As mentioned in chapter 11, mutating genes may capture new code and place those sequences in layaway. And what better place for storage than in the nursery of non-coding DNA? Vital sequences may gather over deep time in the quiet expanses of non-coding DNA. We've also learned that less than 2 percent of our DNA is actively involved in the protein production for our cells and tissues. In the remaining 98 percent, teams of new genes may gather until it's time for them to be expressed. Fredric Menger, professor of organic chemistry at Emory University, labelled this process "preassembly."[269]

---

[269] Fredric M. Menger, "An Alternative Molecular View of Evolution: How DNA was Altered over Geological Time," *Molecules* 25, no. 21 (November 2020): article 5081, https://doi.org/10.3390/molecules25215081.

Question:

Non-coding DNA (previously mislabeled "junk") has been preserved for hundreds of millions of years. Does this longevity suggest that non-coding DNA plays a vital role in the unfolding of the species?

Sequencing of "Living Fossil" Genomes

Confirming the reality of this process and the critical role for preassembly, biologists have sequenced the largest genome of any known species, a "living fossil" called the lungfish. Their conclusion: all the genes required for living on land gathered in the fish genome long before the first frog hopped onto land. (Details are provided in chapter 17.)

Recognizing Life's Nursery

The genome contains vast spaces of non-coding DNA, quiet regions of unused code long considered "junk." Why would life preserve junk for millions of years? For one, it provides a home for regulatory genes, those master genes that control gene expression. Therefore, it makes sense that the unused code serves another vital purpose.[270]

This quiet DNA also serves as a repository for potentially useful adaptations, even entirely new species, allowing teams of mutations to gather until the time for their expression arrives.

---

[270] Jake Buehler, "The Complex Truth About 'Junk DNA'," Quanta Magazine, September 1, 2021, https://www.quantamagazine.org/the-complex-truth-about-junk-dna-20210901/.

New body plans and breakthrough abilities "preassemble" over time like a slow and steady snowfall quietly drifting into the genome. When a complete team of genes for a new appendage or function has gathered and the time for their expression arrives, life flows in a dramatic new direction: Trilobites appear in the ancient seas.[271] The first frog crawls onto land.[272] Whales analyze echoes to identify their favorite meal of giant squid in deep and dark ocean waters.[273] After eons of preparation, human intelligence blossoms around the globe.[274]

Recent research has demonstrated mutational activity in this non-coding DNA, producing just the kind of neutral mutation that could safely visit life's library.[275]

The Maker gives life freedom, enjoying the occasional surprise. Why allow dinosaurs to roam Earth for millions of years? Unaffected by time, God may have enjoyed their stunning diversity. But He ended their reign with an asteroid and opened the world to mammals.[276]

---

[271] Stephen Jay Gould, *Wonderful Life: The Burgess Shale and the Nature of History* (New York: W. W. Norton & Company, 1990), 16.

[272] Bob Strauss, "The 300 Million Years of Amphibian Evolution," ThoughtCo, August 15, 2019, https://www.thoughtco.com/300-million-years-of-amphibian-evolution-1093315.

[273] "In Bats and Whales, Convergence in Echolocation Ability Runs Deep," ScienceDaily, January 27, 2010, https://www.sciencedaily.com/releases/2010/01/100125123219.htm.

[274] Fredric M. Menger, "Molecular Lamarckism: On the Evolution of Human Intelligence," *World Futures* 73, no. 2 (May 2017): 89–103, https://doi.org/10.1080/02604027.2017.1319669.

[275] Li Zhang et al., "Rapid Evolution of Protein Diversity by de novo Origination in *Oryza*," *Nature Ecology and Evolution* 3 (2019): 679–690, https://doi.org/10.1038/s41559-019-0822-5.

[276] Riley Black, *The Last Days of the Dinosaurs: An Asteroid, Extinction, and the Beginning of Our World* (New York: St. Martin's Press, 2022).

Understanding Preassembly

If life tours the mutational landscape and finds a new idea, does it immediately utilize that mutation? Or is there a more promising option, a way to slowly gather hopeful sequences in the genome for use at a future time, like placing a future Christmas gift on long-term hold?

Perhaps neutral genes take advantage of their uncommitted status, entangling with useful sequences stored on the hypercube library and slowly retrieving new code.[277] When an entire team of genes dedicated to a purpose useful to the species has finally gathered, and a signal for their expression arrives, life moves in a new direction. Creatures conquer the land, fly through the skies, or commit to a mate for life. When it comes to macroevolution, this process of gene accumulation provides a vastly superior alternative to Darwin's stutter step "minor variation and selection."

In his essay entitled "An Alternative Molecular View of Evolution: How DNA was Altered over Geological Time," Fredric Menger, the professor mentioned at the start of this chapter, acknowledges that the classical Darwinian model—multiple sequential episodes of mutation and selection—is a good fit for microevolution. Still, it simply fails to explain the abrupt appearance of a major new function. As previously mentioned and noted in the "codes for life," the concept of "preassembly" represents a major contribution to modern evolutionary theory.[278]

As this text and many others have already noted, neo-Darwinism fails to explain the proliferation of new body plans and appendages in the Cambrian seas. Neo-Darwinism also fails to explain the

---

[277] Andreas Wagner, *Arrival of the Fittest: Solving Evolution's Greatest Puzzle* (New York: Current, 2014), 179.
[278] Menger, "An Alternative Molecular View."

abrupt appearance of echolocation in distantly related creatures, discussed in the next chapter.

Alluding to the arrival of seeing with sound in species as diverse as bats and whales, Menger notes: "Although mutations are rare, random, and mainly harmful, they [bats and whales] managed to find their way through a maze of biological complexity, with the aid of natural selection, not once but twice."[279]

How would preassembly explain the sudden appearance of significant innovation? Quiet, non-coding DNA could serve as a "repository," providing a way station for evolutionary developments lying in wait for the future.[280]

Cambrian Explosion Explained

Under preassembly, functional mutations retrieved from the library of life quietly gather in the quiet landscape of non-coding DNA, awaiting the arrival of a complete genetic team. Preassembly allows future genes to hide from the demands of biological function, inactive until life needs their product.

Why did lobsters and insects make a sudden appearance in the Cambrian oceans? The gene sequences for their formation had gathered in preassembly. They existed as non-coding DNA for eons, until some signal—perhaps a change in water temperature or rising oxygen levels—triggered their transcription and translation.

Two facts support a key role for the preassembly of new mutation in so-called junk DNA.

---

279 Menger, "An Alternative Molecular View."
280 Menger, "An Alternative Molecular View."

1. Non-coding, neutral DNA has been preserved over eons in the genome, suggesting it serves a vital purpose.[281]
2. Mutational activity by neutral gene sequences has been carefully documented.[282]

We Have Met the Final Blueprint for Life

The code of complementary base-pairing
The triplet code
The code of protein folds
Regulatory genes
A library of gene and protein sequences
Preassembly

Minor variation one gene at a time guided by natural selection is an important chapter in the story of life, but only reveals part of the story. The preassembly of gene teams dedicated to a singular purpose, biding their time until receiving a signal for transcription, reveals divine purpose and planning.

God's guiding hand is visible once again, after centuries of obfuscation.

Supplement Fifteen

Before life appeared on Earth, preparation was well underway. Small elements such as hydrogen, oxygen, and carbon made major contributions, providing water, the atmosphere, chemical energy, and other life-supporting features.

---

[281] Buehler, "The Complex Truth."
[282] Zhang et al., "Rapid Evolution of Protein."

Review the role of a multi-component protein called ATPase that straddles the mitochondrial membrane (mitochondria are the "batteries" living by the hundreds inside our cells) and rotates 130 revolutions per minute to store chemical energy for the multiple activities of the living cell

How fortuitous that life accidentally "stumbled upon" this amazing nanomachine floating in the hyperspace of proteins.

CHAPTER SIXTEEN

Library in Action

Darwin's chance-based minor variation might impact the color of fur, the length of a beak, or even modify a vital structure that triggers the development of a new species. But a library for life and the preassembly of gene teams provides a better explanation for new body plans, new species, and complex adaptations.

Library Visits

It is a well-established fact of biology that unrelated life-forms often "converge"—employ or return to the same physical features. All fish have tapered bodies, and all birds have feathers. Biologists attribute these similar physical features to the power of natural selection. If minor variation and natural selection are the driving force, genes responsible for this physical convergence should be a motley mix of unrelated sequences, reflecting their origin in a random and accidental process. Has DNA sequencing confirmed the Darwinian model, finding a plethora of unrelated genes accidentally achieving the same function? To the contrary, unrelated animals are finding and utilizing the same genes.

Surviving on a Bamboo Diet

Imagine being a panda and attempting to survive on a bamboo diet. China is home to two pandas: one a black-and-white bear, the other a red-furred cousin of the ferret. Despite sharing a name, the species are unrelated and have never interbred. How do these carnivores survive on a vegetarian diet limited to bamboo? Both species have found identical genes to solve their nutritional challenge.[283]

Echolocation

Imagine needing to see with sound in the darkness of a cave (bats), in deep ocean waters (porpoises and whales), or on the dark forest floor (tree mice). These creatures have developed the same superpower to see with sonar, painting their world with sound.

Bats zip through the nighttime sky using echolocation to capture insects on the fly.[284] At the same time, a sperm whale, one of the most massive creatures on Earth, descends into the darkness of the deep sea and uses echolocation to identify its favorite food: the giant squid.[285] A tree mouse, one of the smallest creatures on Earth,

---

[283] Yibo Hu et al., "Comparative Genomics Reveals Convergent Evolution between the Bamboo-Eating Giant and Red Pandas," *Proceedings of the National Academy of Sciences* 114, no. 5 (January 2017): 1081–1086, https://doi.org/10.1073/pnas.1613870114.

[284] Merlin Tuttle, *The Secret Lives of Bats: My Adventures with the World's Most Misunderstood Mammals* (New York: Mariner Books, 2018); Riley Black, "Why Bats Are One of Evolution's Greatest Puzzles." *Smithsonian Magazine*. April 21, 2020, https://www.smithsonianmag.com/science-nature/bats-evolution-history-180974610/.

[285] Zhen Liu et al., "Genomic and Functional Evidence Reveals Insights into the Origin of Echolocation in Whales," *Science Advances* 4, no. 10 (October 2018): article eaat8821, https://doi.org/10.1126/sciadv.aat8821.

leaves the safety of its tree in the night and uses echolocation to locate nuts on the forest floor.[286]

Somehow these distantly related creatures have acquired similar, if not identical, genes to echolocate, a complex adaptation involving hundreds of genes. Each animal produces sound waves unique to its species. Each has modified their brain to capture and interpret returning echoes. Each navigates the world with sonar.

Darwinists would predict that a random shuffling of genes produced these superpowers, guided by natural selection. Gene sequencing has proven them wrong. The sequencing of genomes from numerous echolocating species has demonstrated that these animals use similar and often identical genes.

Key Definition: Convergence

The process whereby unrelated organisms independently evolve similar traits. Classically, convergence was considered a physical phenomenon imposed by the demands of the environment. Facing similar challenges, natural selection drove animals to similar physical solutions.

Key Definition: DNA Sequencing

Next-generation sequencing refers to the analysis of DNA one letter at a time. Modern technology allows rapid throughput at lower cost. Biologists now recognize that physical convergence reflects a deeper reality: unrelated species often achieve the same look or function by acquiring and expressing the same genes.

---

[286]   Kai He et al., "Echolocation in Soft-Furred Tree Mice," *Science* 372, no. 6548 (June 20210): article eaay1513, https://doi.org/10.1126/science.aay1513.

Convergence through Identical Genes

With the advent of modern sequencing, convergence is taking on a powerful new meaning. The sequencing of genomes has produced astounding and unexpected results. Distantly related species are using—converging on—the same genetic code

As described above, the ability to survive on a diet restricted to bamboo provides an excellent case in point. These are two unrelated species with the same name "panda" that live in the bamboo forests of China. The last common ancestor of these pandas lived forty-three million years ago.[287] The red panda that lives in the eastern Himalayas has no relationship to bears. Its family includes weasels, ferrets, raccoons, and skunks,[288] while the giant black-and-white panda is a bear.[289]

---

[287] Kate Wong, "Meet the Last Common Ancestor of Bats, Whales, Sloths, and Humans," *Scientific American*, February 7, 2013, https://blogs.scientificamerican.com/observations/
meet-the-last-common-ancestor- of-bats-whales-sloths-and-humans/.

[288] John J. Flynn et al., "Whence the Red Panda?" *Molecular Phylogenetics and Evolution* 17, no. 2 (November 2000): 190–199, https://doi.org/10.1006/mpev.2000.0819.

[289] Mingchun Zhang et al., "Giant Panda Foraging and Movement Patterns in Response to Bamboo Shoot Growth," *Environmental Science and Pollution Research* 25 (January 2018): 8636–8643, https://doi.org/10.1007/s11356-017-0919-9.

Two species called "panda" live in the bamboo forests of China. The black and white panda is a bear. The red panda's closest relative is a ferret. Despite having no direct relationship, both species have found similar and often identical genes to survive on a bamboo diet.

Both pandas are members of the mammalian order Carnivora, making it surprising that they are both vegetarians. Both species subsist entirely on bamboo. And both use an extra thumb to manipulate the stalk. Do these unrelated vegetarians owe their peculiarities to changes in their genome? We should let next-generation sequencing give us the answer.

Researchers compared the DNA of the two pandas with a variety of other species, including polar bears (close relatives of giant pandas), ferrets (close relatives of red pandas), and tigers and dogs (close relatives of neither). Have these two panda species who have never interbred found identical genetic solutions to feed on bamboo?

Seventy Genes Changed the Same Way

The team identified seventy genes acquired by both pandas, and another ten genes knocked out by mutation—changes not found in polar bears, ferrets, tigers, or dogs.[290] Besides the genes for the extra thumb, these creatures share other genes that affect the digestion and extraction of nutrients.[291] Their genomes contain identical genes for three digestive enzymes required to extract the essential amino acids lysine and arginine from bamboo proteins. Meat contains an abundance of those amino acids, but bamboo contains meager amounts.

The team also noted similar changes in genes involved in the metabolism of vitamins A and B12 and arachidonic acid, essential molecules for bodily function.

These two species have somehow found identical genetic solutions to survive on a bamboo diet. Neo-Darwinism would praise the power of natural selection: look how selection has achieved this remarkable degree of convergence. But these pandas provide a powerful challenge to the standard model. Unrelated, nonbreeding species have surmounted the same nutritional challenge using identical genes.

---

[290]   Yibo Hu et al., "Comparative Genomics Reveals."
[291]   "Two Strange Mammals Illuminate the Process of Natural Selection," *The Economist*, January 21, 2017, https://www.economist.com/science-and-technology/2017/01/21/two-strange-mammals-illuminate-the-process-of-natural-selection.

Convergent Genes for Echolocation

Dolphins and bats are strikingly different in size, habitat, and means of locomotion. But they share the aforementioned sixth sense, both hunting their prey by emitting high-pitched sounds and analyzing returning echoes to understand their surroundings.[292] Neo-Darwinism would expect the ability to see with sonar to develop through a series of little steps, one component of echolocation at a time.

Unfortunately for neo-Darwinism, there is no evidence of partially echolocating fossils. The last common ancestor of bats and whales—a creature that lived more than sixty million years ago—had no echolocation ability.[293] Even the first ancient bat couldn't echolocate. Yet, this capability appeared fully formed in the fossil record in multiple species of bats and whales. Biologists have even identified one mouse species that exploits nighttime echolocation to find nuts on the forest floor.[294]

Many distantly related animals deep diving in the ocean, flying through the nighttime skies, or foraging on the forest floor have developed this alternative means of sight using sonar. There is no evidence they inherited this superpower.

Is it reasonable to believe that echolocation (or conventional vision with the eye) evolved separately in these animals by natural selection sifting through random mutations? Or was the ability to echolocate created through the preassembly of coordinated code?

---

[292] Fredric M. Menger, "An Alternative Molecular View of Evolution: How DNA was Altered over Geological Time," *Molecules* 25, no. 21 (November 2020): article 5081, https://doi.org/10.3390/molecules25215081.

[293] Sedeer el-Showk, "An Example of Convergent Evolution in Whales and Bats," Accumulating Glitches (blog), Scitable by Nature Education, July 1, 2013, https://www.nature.com/scitable/blog/accumulating-glitches/an_example_of_convergent_evolution/.

[294] Gareth Jones, "Sensory Biology: Tree Mice Use Echolocation," *Current Biology* 31 no. 18 (September 2021): PR1074–PR1076, https://doi.org/10.1016/j.cub.2021.07.074.

Thanks to DNA analysis in the past few years, biologists have answered that question.

One team sequenced the genomes of four species from various branches of the bat family tree, two that echolocate and two that do not. Their study included genomic sequences of the large flying fox and the little brown bat, another echolocator.

These researchers compared bat sequences to those from more than a dozen other mammals. *Their analysis revealed that two hundred genes had independently changed the same way.*[295]

Pay attention to the following quote from a team member (italics mine): *"The biggest surprise...is...the extent to which convergent molecular evolution seems to be widespread in the genome."*[296]

In other words, species that are not that closely related are finding the same genetic solutions. How do these creatures find the same gene sequences? And how are those solutions appearing abruptly in the fossil record? Preassembly is a vastly more powerful explanation than minor variation and selection.

The Same Genetic Coding in Unrelated Species

Bats emit tones from their mouth and nose and analyze returning echoes to capture prey. Each bat species uses a unique range of frequencies, depending on its environment and the target.[297]

Unlike bats, echolocating whales and dolphins produce sound by creating vibrations in a unique organ called the "melon" on the

[295] Joe Parker et al., "Genome-Wide Signatures of Convergent Evolution in Echolocating Mammals," *Nature* 502 (September 2013): 228–231, https://doi.org/10.1038/nature12511.

[296] Elizabeth Pennisi, "Bats and Dolphins Evolved Echolocation in Same Way," *Science*, September 4, 2013, https://www.science.org/content/article/bats-and-dolphins-evolved-echolocation-same-way.

[297] Katy Warner, "Echolocation," National Park Service, October 14, 2020, https://www.nps.gov/subjects/bats/echolocation.htm.

top of their head. These creatures collect the echoes of those sounds through fatty structures in the lower jaw around the ear.[298] The brains of both species have learned to analyze the echoes.

Somehow, these creatures acquired the genes for the structures, proteins, tissue types, cell types, and neural circuits to echolocate— by preassembling hundreds of genes in non-coding DNA.

Dancing Prestin

The protein prestin amplifies incoming soundwaves to deliver more potent messages as they travel to the brain. Prestin plays a vital role in hearing, since it literally "dances" in response to sound, helping hair cells in the inner ear elongate and contract.[299]

In a 2010 study published in the journal *Current Biology*, researchers examined the prestin gene in several dolphin species, the sperm whale, non-echolocating baleen whales, and animals like pigs, cows, and humans. In echolocating animals, fourteen sites along the gene had identical changes.[300] Coincidence? Hardly. Natural selection may have played a role in this convergence, but the question remains: who or what nominated these gene sequences?

Many researchers assume that similarities among organisms in different environments would involve different genes, or at least modified versions of genes, but that is not the case. In an understatement, Gareth Jones, an evolutionary biologist at the University

---

[298] C. J. Harper et al., "Morphology of the Melon and its Tendinous Connections to the Facial Muscles in Bottlenose Dolphins (*Tursiops Truncatus*)," *Journal of Morphology* 269, no. 7 (July 2008): 820–839, https://doi.org/10.1002/jmor.10628.

[299] Jing Zheng et al., "Prestin is the Motor Protein of Cochlear Outer Hair Cells," *Nature* 405 (May 2000): 149–155, https://doi.org/10.1038/35012009.

[300] Yang Liu et al., "Convergent Sequence Evolution between Echolocating Bats and Dolphins," *Current Biology* 20, no. 2 (January 2010): PR53–R54, https://doi.org/10.1016/j.cub.2009.11.058.

of Bristol, stated that the findings in the *Current Biology* study were "very unexpected."[301]

Genetic Convergence

A review in ScienceDaily.com came to this conclusion (italics mine): "*The discovery represents an unprecedented example of adaptive sequence convergence between two highly divergent groups and suggests that such convergence at the sequence level might be more common than scientists had suspected.*"[302]

Translation: Distantly related species use similar mutations for an incredibly complex adaptation—echolocation. While natural selection might play a role, do we really believe a random process could fully explain these findings? Or did the genomes of these creatures visit the same sequence library?

The Next Chapter Will Tell the Tale

A sequencing lab analyzed the largest genome of any known species, a "living fossil" called the lungfish. Wait until you hear how many genes for the future gathered in this ancient fish DNA.

Supplement Sixteen

To acquire gene sequences for future adaptations, mutations in non-coding DNA must visit the sequence library. If those visits

---

[301] Elizabeth Pennisi, "Hear That? Bats and Whales Share Sonar Protein," *Science*, January 25, 2010, https://www.science.org/content/article/hear-bats-and-whales-share-sonar-protein.

[302] "In Bats and Whales, Convergence in Echolocation Ability Runs Deep," ScienceDaily, January 27, 2010, https://www.sciencedaily.com/releases/2010/01/100125123219.htm.

167

involved essential protein-coding genes, the resulting mutation could be harmful. Mutations in non-coding DNA would be "neutral," perfect candidates for library visits.

Geneticists have recently documented the appearance of neutral mutations in just the region of the genome that might participate in preassembly.

CHAPTER SEVENTEEN

Land Ho!

Prepare to occupy the land.

In the preceding chapters, we encountered numerous examples of unrelated species finding identical genes to solve problems or develop new capabilities. These findings challenge the Darwinian mantra that "random variations" weave the story of life.

This chapter approaches the same issue from a different angle. One of life's most important events happened in the ancient seas: the fins of the fish became the limbs of the frog, and the first amphibian hopped onto land.[303] There are two stars in this fin-to-limb movie: *Hox* genes (master control or regulatory genes introduced in chapter 13) made critical changes to the fin of a special fish, the lungfish.[304] Second, the "living fossil" lungfish fish delivered those *Hox* genes to the future.[305]

Like the blueprints of the universe, *Hox* genes execute divine intentions. They shape eyes, jaws, and, yes, limbs. By exploring *Hox*

---

[303] Robert L. Carroll, *Vertebrate Paleontology and Evolution* (New York: W. H. Freeman, 1988).

[304] K. Wang et al., "African Lungfish Genome Sheds Light on the Vertebrate Water-to-Land Transition," *Cell* 184, no. 5 (March 2021): 1362–1376.e1318, https://doi.org/10.1016/j.cell.2021.01.047.

[305] Sahar Knight, "Fish had the Genes to Adapt to Life on Land While They were Still Swimming in the Seas," Front Line Genomics, February 18, 2021, https://frontlinegenomics.com/fish-had-the-genes-to-adapt-to-life-on-land-while-they-were-still-swimming-the-sea/.

gene activity in a fish, we will uncover one of life's great secrets. Did the lungfish make landfall? Not quite, but it made a significant contribution, preparing the way for the first amphibian to arrive on land.

The Fin-to-Limb Transition

One scene in the movie of life deserves top billing: creatures with fins converted those fins into limbs, delivering the first amphibian to dry ground. Somehow, the fin of ancient fish produced the leg of a frog, giving birth to all future species with four limbs (tetrapods). Biologists have long wondered how life moved onto land, developing lungs to breathe, joints to hop, and tissues and organs to survive.[306] Stunning research has clarified the process. The code for living on land gathered in the genome of ancient fish until the time for landfall arrived.[307]

Hox genes conduct the DNA orchestra. The *Hox* toolkit has been carefully preserved for hundreds of millions of years as these overriding genes plug and play, creating one new species after another.[308]

Unchanged for hundreds of million years, the lungfish is a "living fossil."[309] In addition to its air-breathing lung, the lungfish

[306] Axel Meyer et al., "Giant Lungfish Genome Elucidates the Conquest of Land by Vertebrates," *Nature* 590 (January 2021): 284–289, https://doi.org/10.1038/s41586-021-03198-8.

[307] Knight, "Fish had the Genes."

[308] Markus Friedrich, "Evo-Devo Gene Toolkit Update: At Least Seven Pax Transcription Factor Subfamilies in the Last Common Ancestor of Bilaterian Animals," *Evolution & Development* 17, no. 5 (September 2015): 255–257, https://doi.org/10.1111/ede.12137.

[309] Marco Krefting and Rachel More, "Researchers Decode Aussie Lungfish Genome," *Australian Geographic*, January 21, 2021, https://www.australiangeographic.com.au/topics/science-environment/2021/01/researchers-decode-aussie-lungfish-genome/.

has fins like no other fish. The fin of the lungfish contains a radius and ulna, also found in the arm bones of modern humans.[310] And that fin contains *Hox* markers associated with the development of limbs. Every creature with fingers and toes should be grateful for the lungfish fin.

The lungfish possesses the single largest genome of all known species, *forty-one times larger than the human genome.*[311] In 2021, it took a highly sophisticated team over one hundred thousand hours to sequence the lungfish genome, the closest living relative to the first amphibian on land.[312]

The findings? Almost all the sequences required for living on land had gathered in the genome of this ancient fish.

---

[310] M. Brent Hawkins, Katrin Henke, and Matthew P. Harris, "Latent Developmental Potential to Form Limb-Like Skeletal Structures in Zebrafish," *Cell* 184, no. 4 (February 2021): 899–911.e13, https://doi.org/10.1016/j.cell.2021.01.003.

[311] Maria Assunta Biscotti et al., "The Lungfish Transcriptome: A Glimpse into Molecular Evolution Events at the Transition from Water to Land," *Scientific Reports* 6 (February 2016): article 21571, https://doi.org/10.1038/srep21571.

[312] Alex Fox, "Australian Lungfish Has Biggest Genome Ever Sequenced," *Smithsonian Magazine*, January 27, 2021, https://www.smithsonianmag.com/smart-news/australian-lungfish-has-biggest-genome-ever-sequenced-180976837/.

The "living fossil" lungfish has the largest genome of any known species. Biologists succeeded in sequencing the lungfish genome in 2021. What did the sequencing reveal? All the genes required for life on land had preassembled in the lungfish, including the genes for joints, lungs, a larger heart, a radius, and ulna and one of the HOX genes required to produce a limb. Did that gene arrive in its genome as a "minor variation"? Did natural selection play a role in its arrival? Neo-Darwinian theory is facing a crisis.

Question:

Genes for life on land preassembled in the genome of an ancient fish, many of them superfluous to life in the ocean. Does it seem realistic that natural selection was the primary guide in this process?

One Lucky Fish

Lungfish were abundant in the Devonian era (about 380 to 400 million years ago), though only six species survive today.[313] However, modern lungfish appear identical to their ancient fossils, making them a favorite subject of DNA research. Lungfish provide critical information about the genetic changes that enabled life's arrival on land.

Discovered over a century ago, the aptly named lungfish have lungs as well as gills.[314] If oxygen levels drop in the water, this fish can float to the surface and inhale a gulp of air.[315] Lungfish are also characterized by lobed fins. Modern fish have ray fins supported by barbed bones—sharp spikes only too familiar to careless fishermen.[316] The lobed fin of the lungfish is an unimpressive stump, but it contains *Hox* gene treasure.

Darwinists might describe the lungfish as incredibly lucky. As one science journalist expressed it, "If you are a lucky species, you will stumble into random gene mutations that just happen to help you survive better—allowing you and your descendants to keep and build on the helpful traits they encode. As with anything involving luck, the more chances you take, the more chances you have of hitting the jackpot."[317] Darwinists would have us believe

[313] Biscotti et al., "The Lungfish Transcriptome."
[314] G. R. Allen, S.H. Midgley, and M. Allen, *Field Guide to the Freshwater Fishes of Australia* (Perth, Australia: Western Australia Museum, 2002), 54–55.
[315] Gilbert Percy Whitley, *G.P. Whitley's Handbook of Australian Fishes* (Victoria, Australia: Wilke and Company Ltd., 1960), 334.
[316] Lauren C. Sallan, "Major Issues in the Origins of Ray-Finned Fish (*Actinopterygii*) Biodiversity," *Biological Reviews* 89, no. 4 (February 2014): 950–971, https://doi.org/10.1111/brv.12086.
[317] Tessa Koumoundouros, "The Massive Genome of the Lungfish May Explain How We Made the Leap to Land," Science Alert, January 20, 2021, https://www.sciencealert.com/lungfish-s-massive-genome-could-explain-how-we-made-our-leap-from-sea-to-land.

the transition to life on land was a matter of random chance. This chapter will challenge that hypothesis.

Steps on the Journey to Land

The transition to life on land required numerous innovations. In one crucial step, the lungfish supplemented its gills with lungs and acquired a variety of genes to produce surfactants.[318] Surfactants reduce the surface tension of pulmonary air sacs and facilitate lung expansion. Premature infants may spend weeks in the neonatal ICU waiting for their surfactant to mature.

Lungfish also developed mechanisms to avoid ammonia toxicity.[319] The buildup of ammonia is of little concern to the typical fish—they simply excrete nitrogenous waste through their gills. But lungfish have developed the ability to live out of water, burrowing into the mud and "estivating," a form of hibernation when water holes go dry.[320] They coat themselves in a mucous membrane, slow their metabolism by more than 90 percent, and avoid ammonia toxicity by converting ammonia to urea. Urea excretion is widespread among amphibians and mammals, including humans. The switch from ammonia to urea to deal with nitrogenous waste was a crucial step in the journey to land.

Lungfish also developed a new repertoire of receptors to detect land-based odors[321] and genes to regulate emotions, including mutations that promote arousal and moderate the reaction to

---

318 Biscotti et al., "The Lungfish Transcriptome," 6.
319 Biscotti et al., "The Lungfish Transcriptome," 6.
320 R. G. Delaney, S. Lahiri, and A. P. Fishman, "Aestivation of the African Lungfish *Protopterus aethiopicus*: Cardiovascular and Respiratory Functions," *Journal of Experimental Biology* 61, no. 1 (February 1974): 111–128, https://doi.org/10.1242/jeb.61.1.111.
321 Meyer et al., "Giant Lungfish Genome."

stress.[322] But the most significant question remains: what role did the lungfish play in the conversion of fins to limbs? Even more, how did they contribute to the development of fingers and toes?

Exploring Genomes with CRISPR

As noted several times, next-generation sequencing has given biology a powerful new tool, allowing scientists to quickly identify gene sequences in the book of life. Recall that the Human Genome Project involved ten years and three billion dollars to sequence the first human genome.[323] Today, biologists can read a human genome in a single day. Thanks to this enormous progress, biologists have also been able to examine the genome of an ever-expanding list of species.

DNA science has other new tools, as well. Bacteria use a biological tool called CRISPR to slice and dice the DNA of an invading virus. CRISPR finds a string of DNA letters, binds tightly, and makes a precision cut. Biologists have learned to tune CRISPR to any genetic sequence, allowing them to silence or replace entire genes precisely.[324] Both modern DNA sequencing and CRISPR are revolutionizing biological research.

The zebra fish is a modern ray-finned fish and a favorite model for DNA research. Using CRISPR to manipulate zebra fish genes, biologists made a stunning discovery: the genes that produce the leg of the frog are still preserved in zebra fish DNA.

In 2021, Dr. Brent Hawkins, in the Department of Organismic and Evolutionary Biology at Harvard, used CRISPR to

---

[322] Wang et al., "African Lungfish Genome," 1371.

[323] Viktor K. McElheny, *Drawing the Map of Life: Inside the Human Genome Project* (New York: Basic Books, 2012).

[324] Yolanda Ridge, *CRISPR: A Powerful Way to Change DNA* (Toronto, ON: Annik Press, 2020).

inactivate two genes in a zebra fish. Hawkins had difficulty believing the result: the fish developed limb-like structures in its fins.[325] By modifying two proteins, Hawkins unknowingly activated the expression of a *Hox* complex involved in the production of limbs. What happened to the mutated fish? In just a few days, they grew the forelimbs of amphibians in their fins, not just in part but in whole, complete with joints, blood vessels, and muscles.

The master control gene for a frog's leg had been hitching a ride with fish DNA for hundreds of millions of years.

Histological analysis revealed that bones in this frog leg had attachments to muscles, something not seen in the fins of modern-day fish. In today's fish, muscles extend directly from the shoulder to bony fin rays, bypassing bones altogether. Yet, the bones that made a surprise appearance in this mutated fish had potentially functional articulating joints, an absolute requirement for the ability to maneuver on land. This report provides stunning proof that the leg of the frog first emerged in the genome of a fish.

The question remains: what role did the lungfish play?

A Special Role for the Lungfish

As noted above, the lobed fin of the lungfish contains a small radius and ulna that could be destined to become the arm bones of humans. How might these new bones improve the lungfish's well-being? Perhaps they helped the lungfish lift themselves above the water for a gulp of air or stir the ocean floor for food.

The lungfish took one additional step in its fin that provides the focus for this chapter. Lungfish expressed a *Hox* gene in the posterior half of its fin that would someday produce fingers and toes.

---

[325] Hawkins, Henke, and Harris, "Latent Developmental Potential."

Meet Hox13

Researchers have studied lungfish embryos to identify critical steps in the fin-to-limb journey. Genes in the *Hox13* family play an essential role in the development of hands and digits. In one key finding, biologists identified the presence of *Hoxa13*, the *Hox* gene for a primitive hand, in the tiny nub of the lungfish fin just beyond the radius and ulna.[326] But another *Hox13* gene, *hoxd13*, is an absolute requirement for the development of digits.[327]

The lungfish fin has two lobes, a posterior lobe, and an anterior lobe, separated by a longitudinal axis. To visualize the lungfish fin, you might think of a bilobed leaf with a vein down the middle and two leaves on either side. *Hoxd13 was only expressed in the lower half of the fin.* The anterior half remained *Hoxd13* free.[328] Two genes in the anterior half of the limb blocked the *Hoxd13* expression.[329] The lungfish fin would need modification to produce the first frog's fingers and toes.

The lungfish fin took important steps to deliver life to the land, but another creature, perhaps a now-extinct lungfish, would be the first to deliver frogs to the shore.

What steps might complete the transformation? Neil Shubin, PhD, the Robert R. Bensley Distinguished Service Professor of Organismal Biology and Anatomy at the University of Chicago, tackled this mystery. Once again, the *Hox* toolkit lies at the heart of the story. A University of Chicago news article about Shubin and

---

[326] Joost M. Woltering et al., "Sarcopterygian Fin Ontogeny Elucidates the Origin of Hands with Digits," *Science Advances* 6, no. 34 (August 2020): article eabc3510, https://doi.org/10.1126/sciadv.abc3510.

[327] Renata Freitas et al., "*Hoxd13* Contribution to the Evolution of Vertebrate Appendages," *Developmental Cell* 23 no.6 (December 2012): 1219–29, https://doi.org/10.1016/j.devcel.2012.10.015.

[328] Woltering et al., "Sarcopterygian Fin," fig. 4.

[329] Woltering et al., "Sarcopterygian Fin," fig. 4.

his team's findings gives an idea: "[The *Hox* gene mutations created in the study] meant that there were fewer cells to make fin rays, leaving more cells at the fin base to produce cartilage elements."[330] Did the posterior portion of the lungfish fin replace the anterior half of the fin, completing the morphogenetic field for fingers and toes?

This exciting research involving "knock out genes" is ongoing and may continue for years. As coauthor Tetsuya Nakamura explained, "What matters is not what happens when you knock out a single gene, but when you do it in combination…That's when the magic happens." But we should revisit the lungfish fin and reevaluate the stunning revelation it provides for the unfolding of life.

Science may someday complete this story. But we should revisit the lungfish fin. What does it tell us about the plan for life's unfolding?

Walking through a home under construction, you pass through openings designed to support future doors. You take one fact for granted: for every doorframe, the architect had plans for a door. In your mind's eye, you might picture a future door hanging in place, complete with hinges and a doorknob.

The lungfish fin is a metaphorical doorframe. The posterior half-fin expression of *hoxd13* is an essential but incomplete step in assembling a completed door. The posterior half of the fin is a promissory note: there are plans to complete the door. The architect has a complete set of drawings. The fix is in: hands and feet with fingers and toes will someday replace the fin.

330   John Eastson, "Discovery Reveals Evolutionary Path from Fins to Fingers," University of Chicago News, August 18, 2016, https://news.uchicago.edu/story/discovery-reveals-evolutionary-path-fins-fingers.

Were There Advantages to the Expression of Hoxd13 on Half a Fin?

We should ask a vital question about the lungfish fin. Did the expression of *hoxd13* in half of a fin provide an incremental advantage to the lungfish? Did *hoxd13*, nestled in the posterior portion of the fin, make the lungfish more attractive to mates? Did it allow the lungfish to produce more progeny or enjoy a longer life? In each instance, the answer is no. The presence of half steps in the form of *hoxd13* provided no advantage to the fish. But it reflects God's ultimate intention: with just a few more *Hox* additions, life will arrive on the land!

The Will of God

Theologians divide the will of God into three components: they believe God has intentional will, circumstantial will, and ultimate will.[331] God expressed His intentional will in Genesis 1: "'Let the land produce living creatures according to their kinds: the livestock, the creatures that move along the ground, and the wild animals, each according to its kind.' And it was so. God made the wild animals according to their kinds, the livestock according to their kinds, and all the creatures that move along the ground according to their kinds. And God saw that it was good."[332]

In the lungfish, we may be witnessing God's circumstantial will: the *Hox* toolkit was active in the lungfish, preparing the way for hands and feet. We may also see the plan for His ultimate will:

---

[331]   Leslie D. Weatherhead, *The Will of God* (Nashville: Abingdon Press, 1944).
[332]   Genesis 1:24–25 (NIV).

there will be a future creature with an upright stance, their hands free for tools, eager to connect with His spirit.

Have we reached a point where creationists and evolutionists alike can celebrate the role of the lungfish? Was it just a lucky fish? Or was the lungfish a preordained actor in the most epic movie of all—the unfolding miracle of life?

Supplement Seventeen

Ships at sea use sonar to detect underwater threats. So do porpoises and whales. How do we minimize our interaction with these sound-seeing creatures of the sea? The problem is particularly acute as engineers place windmills on the seabed. Their work involves extensive mapping of the ocean floor with sonar to determine windmill locations. Whale beaching on the Northeast coast has reached crisis proportions. In their effort to deal with a climate crisis, is the wind industry creating a crisis for the leviathans of the sea?

CHAPTER EIGHTEEN

Intelligence Rising

Give new species more ability to absorb the world "out there" into the mental world "in here."

A Clear Direction

Among a litany of false assertions, evolutionists have insisted that the unfolding of life had no direction.[333] No matter that single cells became multicellular life-forms. No matter that a lumbering Komodo dragon must soak up the sun before hunting, while a warm-blooded jaguar can hunt night and day. Of course, life had direction, capturing information in ever more capable nervous systems as creatures learned to avoid predators—or to be a predator with ever greater cunning and skill.

We are constantly surprised by the intelligence of our fellow creatures. A small parrot called a conure makes its way around the sofa to nestle against your neck. Deer scatter until they realize you pose no threat, then they calmly consume your flowers. Dogs become beloved family members with uncanny tenderness for human infants.

---

[333] Daniel C. Dennett, *Darwin's Dangerous Idea: Evolution and the Meanings of Life* (New York: Simon and Schuster, 1995).

Life's journey had a clear direction: Build increasingly capable nervous systems. Create a rising tide of intelligence until species become self-aware.

Another favorite trope among misguided evolutionists states that human and chimpanzee DNA is 98.5 percent identical.[334] When they make that misleading statement, they are referring to coding DNA. Many species use similar, if not identical, genes to build the nuts-and-bolts proteins for cells and tissues.

How about differences in non-coding DNA? Sequencing of this section of the genome reveals the stunning truth. Humans have hundreds of regulatory genes dedicated to constructing the brain—completely missing in the chimp.[335]

The chimpanzee's brain stops developing at birth. The human brain matures for more than twenty years.[336]

Still not convinced that rising intelligence was a priority of creation? Meet the octopus. When you look at the octopus, it looks back with a camera eye remarkably like our own. The elaborate neural networks of the octopus are stitched together by the same protocadherin (a type of protein) that wires our brains. The octopus uses tools, mimics hundreds of creatures, and is a camouflage expert. The last common ancestor of the human and the octopus? An ancient creature resembling a clam.[337]

---

[334] "New Genome Comparison Finds Chimps, Humans Very Similar at the DNA Level," National Institutes of Health News, National Human Genome Research Institute, August 31, 2005, https://www.genome.gov/15515096.

[335] Anastasia Levchenko et al., "Human Accelerated Regions and Other Human-Specific Sequence Variations in the Context of Evolution and Their Relevance for Brain Development," *Genome Biol and Evolution* 10, no. 1 (January 2018): 166–188, https://doi.org/10.1093/gbe/evx240.

[336] Tomoko Sakai et al., "Fetal Brain Development in Chimpanzees Versus Humans," *Current Biology* 22, no. 18 (September 2012): R791–R792, https://doi.org/10.1016/j.cub.2012.06.062.

[337] Martin R. Smith and Jean-Bernard Caron, "Primitive Soft-Bodied Cephalopods from the Cambrian," *Nature* 465 (May 2010): 469–472, https://doi.org/10.1038/nature09068.

> **Question:**
> *Alone among tens of millions of species, humans have religious instincts and conscious concern for an eternal future. Could there be a special annex in the sequence library labeled: "Homo sapiens in my image"?*

The Development of Mind

God produced a universe of unimaginable scale and complexity. The English American theoretical and mathematical physicist, mathematician, and statistician Freeman Dyson expressed it well: "The laws of nature, and the initial conditions are such as to make the universe as interesting as possible."[338] God produced a universe brimming with exciting features. He then adorned Earth with life.

How many gene sequences have left their imprint on the flora and fauna of life, creating untold millions of proteins to animate cells? How many metabolic pathways have energized life's journey? Even amid the diversity, there was a commonality. Build neural networks as capable as possible. Let those neural networks feast on every available sensation. Let them produce sight with waves of light and see with echoes in the dark. Let them accomplish hearing through vibrations in their bones and sound waves in the air. Let diverse creatures utilize taste and touch to drink in their world. Build synapses and brain structures that thrive on that sensory awareness. Let those brain structures control their bodies until they articulate intelligible sounds. Let them foster their young and cooperate in groups until they learn to hunt as a team.

---

[338] Freeman J. Dyson, *Infinite in All Directions: Gifford Lectures Given at Aberdeen, Scotland, April–November 1985* (New York: Harper Perennial, 2004).

The journey of life was a workshop specializing in the development of the mind. Divine mind was using matter—to create creatures with more-advanced minds.

There is ample reason to recognize that the development of the mind and its Swiss Army knife of capabilities was a priority of speciation. Trilobites, a resident of the Cambrian seas, came complete with eyes.[339] Jawbones shrank in the earliest mammals and became bones for hearing.[340] Nerve endings in the skin captured information about hot and cold, itch and pain.[341] Multiple senses delivered more and more of the world "out there" to the world "in here" as creatures became more capable of responding to their environment.

Increasingly complex nervous systems accompanied the development of body plans, locomotion, and reproduction. Large brains and cognitive sophistication appeared in multiple species.

Avian Species with Higher Intelligence

Many species have learned to communicate with their peers through howls, chirps, and songs. Numerous animals have learned to make tools and solve problems. More than a few find joy in play and companionship with each other. Eight species, perhaps even more, have passed the self-awareness test, recognizing themselves in a mirror.

Complex brains and high intelligence are the number one evidence for the phenomenon called "convergence," as mentioned

---

[339] Riccardo Levi-Setti, *The Trilobite Book: A Visual Journey* (Chicago: University of Chicago Press, 2014).

[340] Anne Le Maître et al., "Evolution of the Mammalian Ear: An Evolvability Hypothesis," *Evolutionary Biology* 47 (May 2020): 187–192, https://doi.org/10.1007/s11692-020-09502-0.

[341] Yang Liu et al., "VGLUT2-Dependent Glutamate Release from Nociceptors Is Required to Sense Pain and Suppress Itch," *Neuron* 68, no. 3 (November 2010): 543–556, https://doi.org/10.1016/j.neuron.2010.09.008.

in chapter 16.[342] Crows, parrots, dolphins, whales, elephants, primates, and even the fantastic octopus have exhibited multiple features of higher intelligence. High levels of intelligence are invariably bound to complex anatomies in the central nervous system, such as the octopus's vertical lobe containing a donut-shaped brain, the pallium or "bird brain" in avian species, and the primate's cerebral cortex. There is a common theme: high intelligence springs from highly ordered neuronal networks. The more complex the brain, the more likely a species will have high cognitive function.[343]

While we commonly denigrate our fellow humans as "bird brains," the term is ill-advised. Corvids (crows and magpies) possess an intelligence that closely mirrors our own.[344] They use advanced facial recognition, squawking at trainers who have trapped them in the past. And they develop complex solutions for problems. Faced with a nut too hard to crack, crows drop the nut on a highway. Once a car tire breaks the nut, they swoop in for a snack—but *only after a red light has stopped oncoming traffic.*

Crows mate for life and stay together as a family. Older offspring even help raise their younger siblings. Crows also enjoy their playtime. They take apparent pleasure in swooping from the sky to startle an unsuspecting dog.

The African gray parrot outshines the crow. Grays have amassed thousand-word vocabularies and use words in context. An African gray named Alex could name more than fifty objects, distinguish

---

[342] Simon Conway Morris, *Life's Solution: Inevitable Humans in a Lonely Universe* (Cambridge: Cambridge University Press, 2004).

[343] Gerhard Roth and Ursula Dicke, "Evolution of the Brain and Intelligence," *Trends in Cognitive Sciences* 9, no.5 (May 2005): 250–257, https://doi.org/10.1016/j.tics.2005.03.005.

[344] Helmut Prior, Ariane Schwartz, and Onur Güntürkün, "Mirror-Induced Behavior in the Magpie (*Pica pica*): Evidence of Self-Recognition," *PLOS Biology* 6, no. 8: article e202, https://doi.org/10.1371/journal.pbio.0060202.

between colors, count to eight, and understand the concept of same and different.[345]

Like chimps and dolphins, African gray parrots master complex intellectual challenges that stump children younger than six. Grays pass a standard test of intelligence with flying colors: researchers present two equal glasses of juice and pour the contents of one glass into a tall and thin glass, the other into a glass that is short and squat. A child younger than six consistently chooses the taller glass, believing it holds more juice.[346] Not the parrot. Gray parrots realize that the two glasses still contain equal quantities of juice.

Dolphins: The IQ Champions

If you define intelligence as the ability to learn and apply knowledge, think abstractly, and deal with new challenges, dolphins are IQ champions.[347] They learn quickly, solve problems, and demonstrate empathy. Their emotional repertoire includes grief, joy, and playfulness. Spindle neurons in dolphins are associated with advanced cognitive abilities, such as reasoning, communicating, and adapting to change. They have a limbic system—the brain's area devoted to processing emotion—that is even more complex than our own.

Dolphins are born into social networks that can endure for a lifetime. Each dolphin has a signature whistle used to call one

---

[345] Irene M. Pepperberg, *Alex & Me: How a Scientist and a Parrot Discovered a Hidden World of Animal Intelligence—and Formed a Deep Bond in the Process* (New York: Harper Perennial, 2009).

[346] Peter Reuell, "Discerning Bird: Researchers' African Grey Parrot Puts (Young) Humans to Shame in Volume-Focused Tests," The Harvard Gazette, December 15, 2017, https://news.harvard.edu/gazette/story/2017/12/harvard-researchers-test-intelligence-of-african-grey-parrot/.

[347] Janet Mann, ed., *Deep Thinkers: Inside the Minds of Whales, Dolphins, and Porpoises* (Chicago: University of Chicago Press, 2017).

another, and they remember those whistles for decades. They vault into the air in back flips and spins, surf in the wake of boats, and enjoy a game of catch, tossing fish back and forth. They even take turns chasing each other in a game of tag. They quickly learn new skills and share those skills with each other.

One group of bottlenose dolphins in Australia stirs the seabed with a sponge in their mouth, protecting their nose while flushing out fish.

Dolphins watch fellow dolphins and quickly learn from each other. One spent a brief time in a sea park without interacting with its trainers. Upon her release, she tail-walked in the open ocean—not a regular dolphin activity—demonstrating her ability to observe and remember lessons.[348]

The Startling Intelligence of an Octopus

No creature makes a stronger argument for the priority of intelligence than the octopus.[349] Cephalopods functionally are fish, close cousins of the oyster. Our most immediate common ancestor is a slug-like beast that lived 550 million years ago. And yet, they demonstrate intelligence in startling ways, including the use of tools. The octopus, for instance, protects the front of its den with shells and coconut husks before taking a nap.

In captivity, the octopus makes friends with individual keepers and develops hostility toward others, squirting less-favored humans with jets of water. Adept at planning, they turn off lights by spraying water and escape from their cages to eat fish in adjacent

---

348 Ed Yong, "A Once-Captive Dolphin Has Introduced Her Friends to a Silly Trend" The Atlantic, September 5, 2018, https://www.theatlantic.com/science/archive/2018/09/dolphins-tail-walking-trend/569314/.

349 Katherine Harmon Courage, *Octopus! The Most Mysterious Creature in the Sea* (New York: Penguin, 2013).

containers. One octopus climbed out of its cage, walked across a room to a drain opening, and squeezed down a one-hundred-sixty-foot pipe to freedom in the open ocean. They also navigate a maze as well as humans.

Octopuses are masters of disguise, changing the color and texture of their skin. They can impersonate at least fifteen different venomous species to discourage would-be predators.

Cephalopods have a nervous system larger than any other invertebrate, containing more than one hundred million neurons. This substantial number of neurons gives the octopus a highly developed capacity for learning and memory. Like humans, they also have a blood-brain barrier to protect their brains from toxins.[350]

Both humans and the octopus have a camera eye. One can only suspect that there is a master control gene for the eye stored within the octopus genome. There is an obvious resemblance between the retina of the octopus and the human retina, even though an enfolding of skin forms the cephalopod retina, while the human retina is an extension of the brain.[351]

Should we consider the octopus a creature possessing higher consciousness? As pets, they bond with humans and readily learn through imitation. They live in a rich cognitive environment manifested through individual personality traits and even the need to sleep. They play with Lego bricks, and some researchers believe they recognize themselves in a mirror. All this is from an animal whose nearest relative is something like an oyster. Can we imagine a better example of higher intelligence as a divine priority?

---

[350] Xiang Zhang et al., "Transcriptome Analysis of the *Octopus vulgaris* Central Nervous System," *PLoS One* 7, no. 6 (June 2012): article e40320, https://doi.org/10.1371/journal.pone.0040320.
[351] Francisco Javier Carreras, "The Inverted Retina and the Evolution of Vertebrates: an Evo-Devo Perspective," *Annals of Eye Science* 3, no. 3 (March 2018): 1–11, https://doi.org/10.21037/aes.2018.02.03.

Wise Elephants

No review of animal intelligence should ignore the elephant. Elephants can classify humans by their age, voice, and gender. When elephants hear tapes using Maasai warriors' language, a tribe that hunts elephants, the herd huddles together in fear. When they listen to tapes from a non-hunting tribe speaking the same words, the herd stays calm and relaxed.[352]

Elephants use tools. They create fly swatters with shards of grass and plug water holes with chewed bark to prevent evaporation.

Elephants comfort each other when distressed. They mourn the dead, caressing the bones with their trunks and standing near the deceased's body for hours.[353]

Problem-Solving Chimpanzees

Nonhuman primates are the most intelligent mammals of them all.[354] Chimpanzees solve problems, use sign language, and communicate with their trainers using pictorial symbols.

In 2013, members of the Nonhuman Rights Project sued to have Leo and Hercules, two chimps used for scientific studies, be recognized as persons with a legal right to liberty. The court denied the lawsuit, but in April 2015, they pushed for the two chimps to be granted a writ of habeas corpus. This order protects against unlawful and indefinite imprisonment, which many saw

---

[352] "Know Your Enemy," *The Economist*, March 15, 2014, https://www.economist.com/science-and-technology/2014/03/15/know-your-enemy.

[353] Jessica Pierce, "Do Animals Experience Grief?" *Smithsonian Magazine*, August 24, 2018, https://www.smithsonianmag.com/science-nature/do-animals-experience-grief-180970124/.

[354] Jane Goodall, *My Life with the Chimpanzees* (New York: Aladdin Paperbacks, 1996).

as an implicit recognition of the chimps' personhood. It was ultimately not granted, after the judge amended her original order. Finally in 2018, after an agreement between the research center holding them and a chimpanzee sanctuary, trainers moved Hercules, Leo, and seven younger chimps to a chimp sanctuary in Blue Ridge, Georgia.[355]

Chimpanzees work with tools, communicate with complex vocalizations, and are good problem solvers. But as bright as chimps are, their brain power pales compared to our own.

The Evolution of Self-Awareness

Some have argued that evolution is unpredictable, accidental, and direction free. This school, often populated by atheists, describes the process as random or contingent.[356] Others see evolution as predictable, predetermined, and inevitable. This school places its emphasis on convergence.

The convergent school has won this argument, hands down. Both the study of body structure and the sequencing of genes have demonstrated that convergent evolution—the repeated, independent development of similar, if not identical, traits—happens over and over. The study of animal intelligence is the study of convergent evolution.[357]

---

[355] "Two Well-Known Former Research Chimpanzees Relocated to Permanent Sanctuary," Project Chimps, March 21, 2018, https://projectchimps.org/two-well-known-former-research-chimpanzees-relocated-to-permanent-sanctuary/.

[356] Stephen Jay Gould, *The Structure of Evolutionary Theory* (Cambridge, MA: Belknap Press, 2002).

[357] Simon Conway Morris, *The Crucible of Creation: The Burgess Shale and the Rise of Animals* (Oxford: Oxford University Press, 1999).

The evolution of intelligence in multiple independent lineages ranging from crows to parrots, from the octopus to the whale, settles the case.

Life on Earth was destined to become self-aware.

Supplement Eighteen

NASA has landed six rovers on the surface of Mars. Will they find evidence that life once thrived on the red planet?

Toward a New Theory

Divine Guidance is Back

God is back in the story of life with priorities we might have expected. He commanded that the oceans teem and the land bring forth, and life responded to His direction. A remarkable array of species has lived in the seas, crawled onto land, and soared through the skies. Extraordinary creatures have become socially aware and engaged with their world, from corvids and parrots to octopuses, porpoises, elephants, and chimpanzees.

Just as God dispatched timeless blueprints from an eternal library of equations to assemble the cosmos, He also prepared a library for life.[358] During the mutation process, neutral, non-coding genes can visit a vast assembly of gene networks and acquire valuable new sequences for the future.

The genome itself provides a protected space for novel genes to gather. All the genes required for life on land accumulated in the genome of ancient fish. As we will see in the final section, the neural structures necessary for higher consciousness were preloaded

---

[358]   Andreas Wagner, *Arrival of the Fittest: Solving Evolution's Greatest Puzzle* (New York: Current, 2014).

in the genome of *Homo sapiens* before human beings developed spiritual awareness.[359]

God's Time is Not Our Time

The expansion of the universe has occurred over billions of years. So has the development of life. Atheistic skeptics use "over a prolonged period of time" to bludgeon the faithful. Why would God tolerate such a drawn-out process? Genesis tells us that God created the world in six days before resting on the seventh day. Undoubtedly, the skeptics say, the reality of deep time proves that the unfolding of life was a chance-based process. Indeed, they ask, if an omnipotent God had supervised life's journey, why didn't He abbreviate the time required to fulfill His intentions?

The time has come to state the obvious. God is utterly different from human beings. He may live in an eternal now with a detailed view of both the past and the future.

As Moses expresses it in Psalm 90:4, God is free of the limitations imposed by our concept of time:

> "A thousand years in Your sight are like a day that
> has just gone by, or like a watch in the night."[360]

Psalm 102 addresses the same issue:[361]

> In the beginning, you laid the foundations of
> the Earth,
> and the heavens are the work of your hands.

---

[359] Fredric M. Menger, "Molecular Lamarckism: On the Evolution of Human Intelligence," *World Futures* 73, no. 2 (May 2017): 89–103, https://doi.org/10.1080/02604027.2017.1319669.
[360] Psalm 90:4 (NIV).
[361] Psalm 102 (NIV).

> They will perish, but you remain.
> They will all wear out like a garment.
> Like clothing, you will change them,
> and they will be discarded.
> But you remain the same,
> and your years will never end.

We live in a physical world with four known space-time dimensions. As John 4:24 reminds us, God is spirit, free of those encumbrances. Like the libraries He organized before the beginning of time, He occupies a timeless dimension.

God Accomplishes His Intentions One Logical Step at a Time

To repeat a fundamental theme: we should use our understanding of God's action in the universe to inform our understanding of life. In addition to galaxies and stars, God filled His universe with black holes, neutron stars, quasars, asteroids, and other exotic space objects. It took the chemical product of three generations of stars to create our sun and its planets. A short-lived giant star and at least two of its stellar descendants went supernova before our solar system could assemble planet Earth.

The unfolding of life has been equally stunning. Here is a summary of the critical developments in the journey of life.

1. Bacteria

Bacteria produced life-sustaining chemical reactions that animate our cells. Bacteria acquired the genes to extract energy from the environment while growing, reproducing, maintaining biological structures, and responding to changes in the environment.[362] Two of the earliest bacteria became specialists in complementary chemical reactions before merging their talents with complex cells:

i. Specialist in Oxygen-Dependent Respiration: The Mitochondria

A freestanding bacterium learned to break down carbohydrates and fatty acids to generate chemical energy using the flow of hydrogen (protons) through the rotor of a complex protein called ATPase.[363] (Supplement 15 explains this process in detail.)

The cellular respiration consumes molecular oxygen and produces CO_2. This bacterium became mitochondria living by the hundreds in each of our cells, creating chemical energy for life.

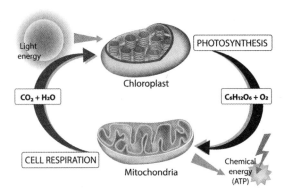

Two complimentary biochemical pathways appeared in early life on planet Earth. During photosynthesis, cells absorb sunlight and CO2 to produce nutrients, secreting oxygen as a byproduct. How fortunate for species like our own that require oxygen to survive.

---

[362] Lynn Margulis, *Symbiotic Planet. A New Look at Evolution* (New York: Basic Books, 1999), 4.

[363] Philip Nelson, *Biological Physics: Energy, Information, Life* (Philadelphia: Chiliagon Science, 2003), 423.

ii. Specialist in Photosynthesis: The Chloroplast

Another ancient microorganism specialized in photosynthesis, combining CO_2 and sunlight to create oxygen and sugar.[364] That bacterium became the chloroplast in plant cells, converting light energy into chemical energy and providing the carbohydrates required for plant growth and human consumption.

One specialized microorganism consumed oxygen and produced CO_2. Another consumed CO_2 and produced oxygen. Is it coincidental that these profoundly reinforcing chemical pathways emerged on early Earth?

2. Blue-Green Algae

Tiny organisms known as blue-green algae specialize in photosynthesis.[365] This little life-form filled the atmosphere with oxygen that sustains us today.

3. Fish

Fish were the first species classified as vertebrates (creatures with a backbone). They have maintained their basic shape for millions of years, surviving mass extinctions and climate change.

4. Amphibians

Amphibians are born in the water and breathe with gills before moving onto land as adults. Amphibians gave us limbs, lungs, more powerful hearts, and other adaptations required for life on land.

---

[364] Margulis, *Symbiotic Planet*, 22.
[365] Robert E. Blankenship, "How Cyanobacteria Went Green," *Science* 355 no. 6332 (March 2017): 1372–1373, https://doi.org/10.1126/science.aam9365.

5. Dinosaurs

Dinosaurs provided breathtaking adaptations with a particular focus on biomechanics, pushing the envelope of physiological possibility. They achieved titanic size, devastating power, razor-sharp teeth, bizarre spines, plates, horns, clubs—and feathers. Following an asteroid impact,[366] dinosaurs morphed into birds.

6. Mammals

Tiny creatures in the treetops threatened by predatory dinosaurs, the first mammals were scarcely larger than a bumblebee and smaller than a squirrel.[367] But mammals acquired traits essential to our future, including an enlarged brain, milk-producing glands, and a diverse array of teeth—incisors, canines, premolars, and molars.

The extinction of the dinosaurs opened the door to mammalian success, providing new habitats and food resources, including flowering plants, insects, fruits, and berries. Mammals entered new kinds of forests, where some became tree-dwelling primates. God set the stage for creatures that walk upright on two legs, their hands free for tools, hunting in groups, and mastering fire.

Intelligence Rising

From bacteria to trilobites, from fish to amphibians, from dinosaurs to mammals, the intentions of God have been abundantly clear: Create an extraordinary planet, bring that planet to life, and

---

[366] Riley Black, *The Last Days of the Dinosaurs: An Asteroid, Extinction, and the Beginning of Our World* (New York: St. Martin's Press, 2022).

[367] John Pickrell, "How the Earliest Mammals Thrived Alongside Dinosaurs," *Nature* (news), October 23, 2019, https://www.nature.com/articles/d41586-019-03170-7.

bring life to self-awareness. Create one unique species with a mind that can grasp God's science and intentions.

Billions of years and millions of species later, we have arrived at His intended destination. God endowed us with His image, including a sense of right and wrong and a hunger for justice and meaning.

A Suggestion for Evolution Deniers

Here's a suggestion for people of faith who recoil at the word "evolution," equating it with a godless worldview: redefine the word in your mind. When you see the word evolution, define it as follows: "the divine unfolding of life." Embrace evolution for what it really represents—a glorious act of creation brimming with free will but accomplishing divine intentions.

The Origin of Life and DNA

We may live in a science-savvy era, but deep mysteries surround us. No scientist has produced a credible account for the origin of life or DNA. Be careful, they warn, don't attribute these wonders to "a god of the gaps." When science finally makes a breakthrough and fills in the gaps, a discredited creator will be once again shown the door.

To repeat a basic premise yet again, I don't see God in "gaps." I see Him in the super brilliance of science we can see and understand.

Creating Human Beings in His Image

This gift of higher consciousness provides a promissory note: God's plan includes something more than a brief lifetime on this Earth.

Something More

We sense "something more" in the mystery of quantum particles: they behave like messengers from another dimension.

We feel that "something more" in the eternal nature of God's blueprints and the creative power of His libraries.

We sense "something more" in the inevitable rise of intelligence in our fellow creatures, in the mental capabilities of an octopus, a grey parrot, or a dolphin.

Our consciousness is a promissory note that we are more than matter in motion. We have an immaterial mind that can lift us above our physical body and give us a glimpse of eternity.

We have toured the creation of the universe and the unfolding of life. The time has come to address the reason for it all. The world became self-aware and God-hungry as God created men and women in His image.

Question:
Do you believe sacred science can tell us about human consciousness and the power and priority of love?

SECTION THREE
Consciousness

THE BLUEPRINTS (ELEMENTS) OF CONSCIOUSNESS

The following is a simple glossary of the blueprints commissioned to create our consciousness, which you will be reading about further in this section.

The Material Brain

The human brain is the most complex form of matter in the universe. Right and left hemispheres are divided into four lobes each. A frontal lobe is responsible for executive action. The occipital lobe accepts visual input. Parietal and temporal lobes are responsible for sensory association, speech, and movement.

The cerebral cortex has rich connections with the thalamus, the central hub of the brain below the cerebral hemispheres. A midbrain and other structures connect the cortex with the spinal cord, while the cerebellum fine-tunes muscle movement and balance. Myelin coats the fibers connecting these components—creating "white matter" that connects the "grey matter" of neurons and distributes information across the brain.

Neural Networks

A neural network is composed of a group of neurons functionally connected by junctions called "synapses," handshakes passing electrical and chemical messages between brain cells. Bushy, tree-like dendrites receive input from axons, the wire-like extension of neighboring neurons. The human brain contains eighty-six billion neurons, and each neuron can construct more than seven thousand synapses. The total synapses in the human brain: *one thousand trillion.*

Electromagnetic Fields

Electromagnetic fields (EM fields) are a field of force produced by the movement of electric charge. One theory of consciousness proposes that consciousness is produced by a superposition of EM fields in the cerebral cortex, delivering components of a mental movie through rich connections to the thalamus at the speed of light. Waveforms on an EEG reflect the electrical activity of the brain.

The Connectome

There is a worldwide effort to complete a map or wiring network of all neural pathways in the brain, including every neuron and synapse. In 2009, the National Institutes of Health announced the "Human Connectome Project," reminiscent of the Human Genome Project, with a goal of creating a nanoscale connectome of the entire human brain. The project will require decades of tedious research but should greatly improve our understanding of brain structure and function.

Mathematics through the Mind's Eye

The ability to manipulate numbers is an innate skill, inherent in humans from birth and enhanced through formal education. The mind's eye is an ability to "see" or imagine with the brain. Great mathematicians believe they make mental visits to a mathematical landscape and use their mind's eye to retrieve beautiful equations.

The Default Mode Network

Distributed components of the brain are devoted to introspection and autobiographical thought, creating a default mode network most active when the brain is at rest. This "daydreaming" network, also described as the "ego" network, is implicated in rumination and depression. The default mode network can hijack the brain and create an unhealthy focus on self.

Meditation and Psychedelic Receptors

The neurotransmitter serotonin helps neurons stabilize mood and creates a sense of happiness and well-being. Serotonin binds to the neuron membrane through a family of receptors, one of which is known as 5-HT2A. Serotonin is unable to penetrate the neuron membrane, but the psychedelics LSD, psilocybin, and DMT, the active ingredient in ayahuasca, readily enter the cell and bind to multiple 5-HT2A receptors throughout the neuron, often producing a mystical experience.

Neuroimaging of subjects reaching a "peak" during meditation demonstrates reduced activity in the default mode network, compatible with temporary dissolution of the ego. Treatment with psychedelics has the same effect.

Love

Love is a force of nature and a growth factor for human development. Orphans deprived of loving care have underdeveloped brains and lifelong difficulty bonding with other people. Individuals treated with psilocybin encounter cosmic love. In addition to inducing a mystical experience, treatment with psilocybin promotes the production of dendrites and their synaptic connections. Love is "neuroplastic," a growth factor for the human brain. Love enhances brain development, worldviews, relationships, and forms the basis for our connection with God.

CHAPTER NINETEEN

The Most Complex Object in the Universe

The brain partners with the mind to produce the story of our lives, complete with subjective experience, memories, thoughts, opinions, and dreams.

Contrasting Views

As we should expect from our tour of the universe and the journey of life, there are divergent views regarding the source and significance of human consciousness.[368] Those who dismiss the wonders of the cosmos as "a few fields and forces" and life as a product of random chance have minimal appetite for a divinely sourced mind, much less an eternal soul. Material skeptics dismiss consciousness as a secondary product of the material brain, just an emergent product of neurons promoting our short-term survival.[369]

A variety of theories compete to explain consciousness. In the material worldview, our thoughts and emotions and memories—what philosophers call our "qualia," the tastes, colors, touch,

---

[368] Rita Carter, *Exploring Consciousness* (Berkeley: University of California Press, 2002).

[369] Daniel C. Dennett, *Consciousness Explained* (New York: Back Bay Books, 1992).

pain, and other components of our rich inner life—emerge from the interaction of electrons and neurochemicals flowing through the brain. To summarize their perspective: "Minds are simply what brains do."[370] Can the mind extend beyond the confines of the skull through telepathy or other means? Materialists dismiss the concept, despite decades of fascinating research. Are "near-death experiences" real? Materialists likely think, "You must be kidding!"

At the same time, a thoughtful group of philosophers and neuroscientists have an entirely different view. Bernardo Kastrup, a leading spokesperson for a worldview known as "idealism" and author of numerous books, including *Why Materialism is Baloney*, expresses it as follows: "A tremendous mystery unfolds in front of our senses every waking hour of our lives; a mystery more profound, more tantalizing, more penetrating and urgent than any novel or thriller. This unfolding mystery is nature's challenge to us. Are we paying enough attention to it?"[371]

Kastrup summarizes the argument from the well-respected Stanford physicist Andrei Linde, father of the theory of cosmic inflation following the big bang: "Matter is an inference and mind a given."[372]

Idealists view the brain as the metaphorical equivalent of a cellphone, receiving input from taste, touch, smell, sight, and sound to produce a multimedia movie of the mind. When material skeptics remind us that a blow to the brain can put an end to conscious life, proving that the mind is a fragile product of the brain, idealists respond in turn: Deliver the same blow to your cellphone.

---

[370] Marvin Minsky, *The Society of Mind* (New York: Simon and Schuster, 1988), 287.

[371] Bernardo Kastrup, *More Than Allegory: On Religious Myth, Truth and Belief* (Summit, PA: Iff Books, 2016).

[372] Bernardo Kastrup, *The Idea of the World: A Multi-Disciplinary Argument for the Mental Nature of Reality* (Summit, PA: Iff Books, 2019), 33.

The voices and the music will stop, but individuals and streaming services on the calling side will remain very much alive.

Materialists believe "minds are simply what brains do." Idealists believe the brain provides temporary housing for an immaterial and personalized mind.

There is yet another way for us to discern the relative merits of these two worldviews. Materialists see the mind as an incidental effect of matter that will die with the death of the brain. Idealists (including most people of faith) see the mind as the currency of creation, as real as space-time or electromagnetic fields, personalized and filtered by the brain en route to life eternal. How do our tours of the universe and life help us sort through these options?

Matter is a Derivative Product

Mathematical equations form the foundation for our universe. Somehow God converted those equations into the fields and forces (blueprints) of physics that have printed out a universe that challenges our comprehension. It should come as no surprise that math provides an unreasonably effective tool for exploring the natural sciences. God is the supreme mathematician. Equations came first. The development of matter required:

1. That light give birth to quarks
2. That the Higgs field encode those quarks with mass
3. That the strong force wraps up and down quarks into three-quark teams to complete the construction of protons and neutrons, the first form of matter

The conclusion is unavoidable: *mind is the master of matter, and matter is a derivative product.*

The equations of the universe are stored in a mathematical landscape in another dimension. So are the sequences for life.[373] As a reminder, mutating genes in the vast stretches of non-coding DNA visited life's library and assembled the genes for life on land in the genome of an ancient fish.[374]

In a final chapter of creative genius, mathematical sequences encoded the genes for neural structure, producing the most complex material object in the universe. Untold trillions of neural maps provide the scaffolding for the human mind.[375]

Would the Architect of the universe—the Creator of life and the source of the mind—subjugate human minds to the electromagnetic fields and neurochemicals of a short-lived material brain? Or would He share His mind with human beings at the beginning of a never-ending story?

Where does sacred science lead us? Do we see the mind as "what the brain does" during a short-term life on Earth? Or do we celebrate the mind as the source of creation that delivers a powerful promissory note: we are both body and soul, destined for life eternal.

---

[373] Andreas Wagner, *Arrival of the Fittest: Solving Evolution's Greatest Puzzle* (New York: Current, 2014).

[374] K. Wang et al., "African Lungfish Genome Sheds Light on the Vertebrate Water-to-Land Transition," *Cell* 184, no. 5 (March 2021): 1362–1376.e1318, https://doi.org/10.1016/j.cell.2021.01.047.

[375] Carl Zimmer, "100 Trillion Connections: New Efforts Probe and Map the Brain's Detailed Architecture," *Scientific American*, January 1, 2011, https://www.scientificamerican.com/article/100-trillion-connections/.

> **Question:**
>
> *Subjective conscious experience is the most real and human thing we know, and yet it remains a mystery. Is the mind a transient product of a material brain or a God-given gift that should shape our understanding of the world?*

Examining Worldviews

How do we evaluate these two divergent views, one seeing the mind as a short-term survival package, ginned up by the material structure of the brain, and the other recognizing mind as the purpose of creation, as central to our existence as the fields of physics and the sequences of DNA?

We will begin in this chapter with a brief but critical tour of the material brain—its large-scale lobes and structures, its neurons, synapses, and neurochemicals, and most importantly, its connections that convert near-innumerable neural maps into a connectome more expansive than our genes (the genome) and the hyperspace of potential proteins (the proteome).[376] We can only marvel at this electrochemical machine that consumes 20 percent of our waking calories as it energizes our mental life.

Then, later chapters will examine the phenomenon called the mind and ask the following questions:

1.   Can the mind extend beyond the brain?

     Since we know gifted mathematicians use their "mind's eye" to discover the equations of black holes and identify

---

[376]   Human Connectome Project, NIH Blueprint for Neuroscience Research, https://www.humanconnectome.org/.

quarks sequestered in the deepest recesses of the atom,[377] is it then so far-fetched to believe that an individual mind can connect with other minds (telepathy), see into the future (precognition), or view objects at a distant location (clairvoyance)? Have materialists unfairly dismissed the scientific study of "psi"?[378]

2. Can the mind function despite a damaged brain?

    a. Children with advanced hydrocephalus, a condition in which high-pressure spinal fluid compresses the physical brain to a small percentage of its normal size, can have a normal IQ.[379]

    b. Severely demented individuals near comatose for years can experience a return to normal speech and mental clarity hours and days before death.[380] Does this "paradoxical or terminal lucidity" suggest that an intact, indestructible mind is escaping the prison of a deteriorating brain?

    c. Thousands of individuals describe a near-death experience (NDE) that can include encounters with the predeceased and a detailed life review. Would we

---

[377] Patchen Barss, "How a Silence Solved the Weird Maths Inside Black Holes," BBC, October 9, 2020, www.bbc.com/future/article/20201008-the-weird-mathematics-that-explains-black-holes-exist.

[378] Thomas Rabeyron, "Why Most Research Findings About Psi Are False: The Replicability Crisis, the Psi Paradox and the Myth of Sisyphus," *Frontiers in Psychology* 11 (September 2020): article 562992, https://doi.org/10.3389/fpsyg.2020.562992.

[379] Roger Lewin, "Is Your Brain Really Necessary" *Science* 210, no. 4475 (December 1980): 1232–1234, https://doi.org/10.1126/science.7434023.

[380] Alex Godfrey, "'The Clouds Cleared': What Terminal Lucidity Teaches Us About Life, Death and Dementia," *The Guardian*, February 23, 2021, https://www.theguardian.com/society/2021/feb/23/the-clouds-cleared-what-terminal-lucidity-teaches-us-about-life-death-and-dementia.

expect these vivid mental experiences to occur when the material brain is barely functioning?[381]

3. Is there evidence that the mind is dependent on love and the influence of religion?

    Neglected Romanian orphans deprived of love suffer grossly abnormal brain development and lifelong difficulty bonding with other humans.[382] Youth involved in religious activities display lower drug abuse and more involvement with their community.[383] Adults involved in religion enjoy longer and healthier lives.[384] Is there a "Holy Spirit field" promoting the power of love?

4. Is there evidence that the mind is eternal?

    As noted above, individuals experiencing an NDE often describe a detailed life review. Does each of us leave a detailed record of our life in a memory bank of the cosmos? Survivors of the NDE exhibit lasting personality changes, including greater altruism and compassion, a sense of mission, and significantly reduced fear of death.[385]

---

[381] Bruce Greyson, *After: A Doctor Explores What Near-Death Experiences Reveal about Life and Beyond* (New York: St. Martin Essentials, 2021).

[382] Shaun Walker, "Thirty Years On, Will the Guilty Pay for Horror of Ceauşescu Orphanages?" *The Guardian*, December 15, 2019, https://www.theguardian.com/world/2019/dec/15/romania-orphanage-child-abusers-may-face-justice-30-years-on.

[383] F. F. Marsiglia et al., "God Forbid! Substance Use Among Religious and Nonreligious Youth," *American Journal of Orthopsychiatry* 75, no. 4 (2005): 585–598, https://doi.org/10.1037/0002-9432.75.4.585.

[384] Marino A. Bruce et al., "Church Attendance, Allostatic Load and Mortality in Middle Aged Adults," *PLOS One* 12, no. 5 (May 2017): article e0177618, https://doi.org/10.1371/journal.pone.0177618.

[385] Greyson, *After*, 88.

The default mode network[386] in the human brain, also known as the "ego" network, maintains our focus on the here and now, thus promoting our day-by-day survival. Meditation and psychedelics reduce the activity of the default mode network, allowing some individuals to experience a "cosmic consciousness" where all is one, and love is the currency of creation. Following treatment with psilocybin, the active component of magic mushrooms, six out of ten atheists drop the atheist label, according to studies at Johns Hopkins University.[387]

Together, the NDE and the life review, terminal lucidity, and psychedelic-induced exposure to the mind at large share a central theme: the material brain provides a temporary home for minds that survive physical existence.

Complexity in Your Hands

Should you ever hold a human brain in your hand, you would immediately appreciate its complexity.[388] There are right and left hemispheres, each divided into four lobes. The frontal lobe is involved in self-awareness, judgment, and concentration. Patients who suffer damage to the frontal lobe can exhibit marked personality changes.

The parietal lobe processes sensory input, including touch, pain, and spatial orientation. The occipital lobe in the back of

---

[386] Andreas Horn et al., "The Structural-Functional Connectome and the Default Mode Network of the Human Brain," *NeuroImage* 102, pt. 1 (November 2014): 142–151, https://doi.org/10.1016/j.neuroimage.2013.09.069.

[387] R. R. Griffiths et al., "Psilocybin can Occasion Mystical-Type Experiences having Substantial and Sustained Personal Meaning and Spiritual Significance," *Psychopharmacology* 187 (2006): 268–83, https://doi.org/10.1007/s00213-006-0457-5.

[388] Rita Carter, *The Human Brain Book: An Illustrated Guide to its Structure, Function, and Disorders* (London, UK: DK Publishing, 2019).

the brain processes the visual information arriving from the optic nerve. Neuronal networks tease the data apart as one group of neurons recognizes vertical lines, another identifies horizontal lines, while others respond to color. That sliced-and-diced information moves on to other brain centers that somehow compose an integrated image. The temporal lobe provides the same capability for the dissection and reconstruction of language.

Supporting cells called "glia" outnumber neurons ten to one, while an abundance of white matter connects neurons and neural networks. The processing power of the brain is most apparent in the second and third layers of the brain's outer cortex—layers of the cortex bristling with enough dendritic branches to carry out high-order functions such as thought and planning.[389]

The Human Connectome

Neuroscientists are hard at work attempting to map these brains connections. As technology advances, they hope someday to map the entire human connectome.[390] But even with a full diagram of connectome wiring, one challenge may remain:

How do these neural nets produce the multimodal mental movie we call the mind?

---

[389] Albert Gidon et al., "Dendritic Action Potentials and Computation in Human Layer 2/3 Cortical Neurons," *Science* 367, no. 6473 (January 2020): 83–87, https://doi.org/10.1126/science.aax6239.
[390] Zimmer, "100 Trillion Connections."

Key Definition: Synapse

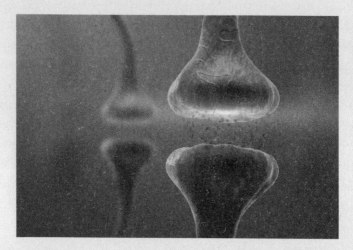

Eighty-six billion neurons produce over one hundred trillion synapses in the human brain. A variety of "neural networks" control our emotions and produce our thoughts. The effects of psychedelics such as psilocybin are producing new insight into our mental function.

The handshake between dendrites and axons, delivering the flow of information across the brain. Neurochemicals such as serotonin and dopamine impact synaptic function.

Regulatory Genes for the Brain

As previously noted, outdated scientists continue to remind us that humans and chimps have nearly identical DNA.[391] This misleading comparison refers to coding DNA, the small fraction of our genome that supervises the construction of proteins supporting cells and tissues. Those genes are strikingly similar in species as diverse as mice and humans. Scientists empowered by next-generation sequencing have moved beyond a fixation on protein-coding genes to explore more extensive regions of DNA. What have they discovered?

Once considered "junk," non-coding DNA houses hundreds, if not thousands, of regulatory genes that contribute to the development of the human brain. These genes are nonexistent in the chimp. What do these "human accelerated regions" do? They provide vital regulatory oversight for human brain development.[392]

Promoting the Growth of Neurons

One trio of genes, known by the scientific name *NOTCH2NL*, allows neuronal stem cells to produce multiple copies of themselves.[393] (This gene is missing in the genome of apes; humans have three copies.) These genes and others yet to be identified build the foundation for intelligence, a vastly complicated phenomenon that separates humans from all other creatures on Earth.

[391] "New Genome Comparison Finds Chimps, Humans Very Similar at the DNA Level," National Institutes of Health News, National Human Genome Research Institute, August 31, 2005, https://www.genome.gov/15515096.

[392] Anastasia Levchenko et al., "Human Accelerated Regions and Other Human-Specific Sequence Variations in the Context of Evolution and Their Relevance for Brain Development," *Genome Biol and Evolution* 10, no. 1 (January 2018): 166–188, https://doi.org/10.1093/gbe/evx240.

[393] Ikuo K. Suzuki et al., "Human-Specific *NOTCH2NL* Genes Expand Cortical Neurogenesis through Delta/Notch Regulation," *Cell* 173, no. 6 (May 2018): 1370–1384.e1316, https://doi.org/10.1016/j.cell.2018.03.067.

Neurons conduct their operations via the flow of ions and neurochemicals through connections with neighbors called "synapses."[394] The neurons have bushy "dendrites" on one end and a long, slender fiber called an "axon" on the other. By establishing relationships with multiple neurons, neurons build a jigsaw puzzle that defies our comprehension. How many synapses do we have in our brain? Over one thousand *trillion*.[395] There are many more synapses in the human brain than there are stars in the Milky Way or atoms in the universe.

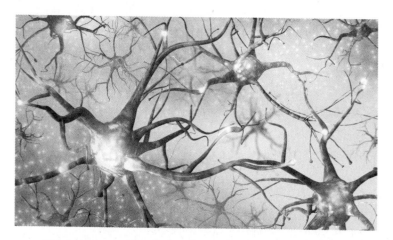

Our neurons communicate with each other through trillions of synapses that transmit information through neural networks, producing subjective conscious experience. Multiple neurotransmitters modulate synaptic function.

---

[394] A. L. Hodgkin and A. F. Huxley, "A Quantitative Description of Membrane Current and its Application to Conduction and Excitation in Nerve," *The Journal of Physiology* 117, no. 4 (August 1952): 500–544, https://doi. org/10.1113/jphysiol.1952.sp004764.

[395] "Neurotransmission: The Synapse" Dana Foundation, September 21, 2023, https://dana.org/resources/neurotransmission-the-synapse/.

Brain Imaging and "Neural Correlates": The Easy Problem

Science has explored the activity of the brain with a variety of imaging techniques. Powerful new brain-scanning machines take images of the brain in living color, determining second-by-second changes in blood flow. Neuroradiologists use these tools to identify "neural correlates" of the consciousness.[396] One area of the brain processes visual information, while another controls the movement of muscles, and yet another facilitates the ability to speak.

The Hard Problem

Critics believe the search for neural correlates is addressing the "easy problem" of consciousness: the association between certain brain areas and individual features of sensation, speech, or movement. These images fall far short of explaining consciousness.

No scan of the brain has yet come close to explaining subjective experience. The supplement to this chapter reviews the leading theories.

Still Hopeless

A brief visit to a dog park confirms that our best friends have a rich emotional life.[397] At one moment, they might seem overwhelmed by their canine companions. Seconds later, they are joyously play-fighting. They are also masters of emoting at home. *Is it time*

---

[396] Geraint Rees, Gabriel Kreiman, and Christof Koch, "Neural Correlates of Consciousness in Humans," *Nature Reviews Neuroscience* 3 (April 2002): 261–270, https://doi.org/10.1038/nrn783.

[397] McConnell, Patricia, *For the Love of a Dog: Understanding Emotion in You and Your Best Friend* (New York: Ballantine Books, 2007).

for a treat? Can we go for a walk? You've been gone fifteen minutes—it seemed like hours! We should celebrate your homecoming with a walk or a treat.

Thomas Nagel, the author of one of the most provocative essays dealing with consciousness, would argue that we can never know what it is like to be a dog.[398] We might imagine living in a pack, sleeping in a cage, greeting each other with smells, and playful posturing. But we will never know what it is like for any individual dog to be a dog.

Imagine how much more complex the subjective flow of consciousness is for a human being. Unlike our four-legged friends, our flow of thought and emotion can contemplate ancient history and imagine the far-distant future.

The mind-body problem reigns supreme among ongoing challenges to the material worldview. As Nagel sarcastically contends: "Without consciousness the mind-body problem would be much less interesting. With consciousness it seems hopeless."[399]

The Reason We are Here

No matter its origin, leading philosophers see consciousness as the primordial substance, an extension of the mind that planned and actualized creation. Oxford mathematician Sir Roger Penrose suggests that "our consciousness is the reason the universe is here."[400] A spiritual approach to consciousness recognizes that the mind has been in the universe from its inception. Consciousness might be

---

[398] Thomas Nagel, "What Is It Like to Be a Bat?" *The Philosophical Review* 83, no. 4 (October 1974): 435–450, https://doi.org/10.2307/2183914.

[399] Nagel, "What Is It Like to Be a Bat?"

[400] Steve Paulson, "Roger Penrose On Why Consciousness Does Not Compute," Nautilus, April 27, 2017, https://nautil.us/roger-penrose-on-why-consciousness-does-not-compute-236591/.

the source of all there is, or the "ground of being." The philosopher Paul Tillich would agree with this view. He saw God as the ground upon which all beings exist.[401]

In a Single Chapter, We Have Met Multiple Elements of Human Consciousness

The brain
Neural networks
Electromagnetic fields
The connectome

Supplement Nineteen

This supplement reviews the leading theories of consciousness. Which one seems closer to the mark: integrated information theory, the global neuronal workspace theory, or theories involving quantum fields? Are they mutually exclusive? Do any one of these theories provide for the possibility that "mind stuff" has an existence independent of the material brain?

As one neuroscientist has expressed it, "Theories of consciousness have excluded consciousness itself." He continues, "Neuroscientists tend to see consciousness as some sort of minimal subsystem of the brain, possessing no information, almost useless. The steam from an engine."[402]

The "hard problem" remains: what produces the individual components of conscious experience, the individual pixels of taste,

---

401   Paul Tillich, *Biblical Religion and the Search for Ultimate Reality* (Chicago: The University of Chicago Press, 1964), 82.
402   Erik Hoel, *The World Behind the World: Consciousness, Free Will, and the Limits of Science* (New York: Avid Reader Press, 2023).

touch, or emotion? These individual ingredients, called "qualia," include colors, tastes, and textures, even memories and thoughts. How does a colorless hunk of matter looking very much like a slab of swiss cheese produce the technicolor movie of our mind?

CHAPTER TWENTY

Mathematics

Mathematics reflects the nature of God.

The Maker of the Heavens and the Earth depended on the individual and collective abilities of mathematically based blueprints to guide the unfolding of the universe. We should repeat the fundamental theme of this book. Where do those blueprints spend eternity? The Higgs field was part and parcel of the big bang package and an absolute requirement for a physical universe.[403] Peter Higgs discovered this field by reviewing a mathematical equation. Where did that equation exist before the creation of the world?

Equations that describe and animate our universe are ancient and eternal. The minds of great mathematicians can *see beyond the confines of their material brain* to capture and retrieve those truths. As previously stated, leading mathematicians believe they retrieve their equations from a landscape or library residing in a higher dimension.[404]

Sometimes mathematical discoveries have immediate application. At other times, a branch of mathematics seems like an irrelevant game, only to provide perfect solutions to scientific puzzles in

---

[403] Frank Wilczek, "A Particle that May Fill 'Empty' Space," *The Wall Street Journal*, December 29, 2022, https://www.wsj.com/articles/a-particle-that-may-fill-empty-space-11672337133.

[404] Roger Penrose, *The Road to Reality: A Complete Guide to the Laws of the Universe* (New York: Vintage Books, 2004).

the near or distant future. It is one thing for mathematicians to discover breakthrough equations to explain an existing force of nature. It is something else for mathematicians to develop equations with no apparent application—only to recognize years later that those equations address fundamental aspects of reality.[405]

In the 1850s, Bernhard Riemann developed a new type of geometry describing curved surfaces, like a sphere or a saddle (in contrast to the geometry of flat surfaces studied in school today).[406] Sixty years later, Einstein realized that Riemann's geometry provided a long-sought foundation for his general theory of relativity. Einstein's theory explains gravity as the product of space-time curvature—a geometry perfectly described by Riemann.

In 1798, Napoleon appointed Joseph Fourier secretary of the Cairo Institute. During his time in Egypt, he pursued interests in mathematics, developing a tool that breaks waveforms into sines and cosines. Are "Fourier transforms" relevant in the modern day? Yes. They form the basis for CT scanning and audio processing.[407] As you enjoy your favorite playlist, think of Joseph Fourier.

Thoughtful scientists marvel at "the unreasonable effectiveness of mathematics in the natural sciences."[408] That effectiveness suggests a two-part revelation: the equations of the universe and the codes of DNA originated in a superintelligent mind. We can retrieve and understand these equations because God created human minds in His image.

---

[405] Paul Davies, *The Mind of God: The Scientific Basis for a Rational World* (New York: Simon and Schuster, 1993).

[406] Luther Pfahler Eisenhart, *Riemannian Geometry* (Princeton, NJ: Princeton University Press, 1997).

[407] Ronald Bracewell, *The Fourier Transform and Its Applications*, 3rd ed. (New York: McGraw-Hill, 1999).

[408] Eugene P. Wigner, "The Unreasonable Effectiveness of Mathematics in the Natural Sciences," *Communications on Pure and Applied Mathematics* 13, no. 1 (1960): 1–14, https://doi.org/10.1002/cpa.3160130102.

Question:
Mathematical equations describe every component of the universe. Is it likely a mindless nature produced the equations of physics and the codes of DNA?

Math and the Search for Meaning

In 1959, Eugene Wigner, a Hungarian American theoretical physicist who received the Nobel Prize in Physics in 1963 "for his contributions to the theory of the atomic nucleus and the elementary particles," paid tribute to the power of mathematics: "The miracle of the appropriateness of the language of mathematics to the formulation of the laws of physics is a wonderful gift which we neither understand nor deserve."[409] This statement addresses the uncanny ability of mathematics to provide insight into the structure and function of the material world. It also refers to the power of math to provide breadcrumbs for discoveries in the future.

In the 1800s, James Clerk Maxwell provided the theoretical foundation for the physics of electromagnetism by showing that electricity and magnetism are aspects of the same field.[410] A century later, in the 1970s, scientists used Maxwell's mathematics to prove that electromagnetism is part of a larger force called the "electroweak force."[411] Scientists also updated our understanding of electricity and magnetism through the mathematics of quantum

---

[409] Wigner, "Unreasonable Effectiveness."

[410] James Clerk Maxwell, "A Dynamical Theory of the Electromagnetic Field," *Philosophical Transactions of the Royal Society of London* 155 (December 1865): 459–512, https://doi.org/10.1098/rstl.1865.0008.

[411] A. Salam and J. C. Ward, "Electromagnetic and Weak Interactions," *Physics Letters* 13 no. 2 (November 1964): 168–171, https://doi.org/10.1016/0031-9163(64)90711-5.

theory. Step-by-step, scientists uncovered the math to arrive at a complete account of the interaction between light and matter. The great quantum physicist Richard Feynman called quantum electro-dynamics—the fully developed theory of the interaction of light with matter—"the jewel of physics."[412] The math underlying that "jewel" existed before the first ray of light.

Ancient Math is Still True Today

Born in 570 B.C., Pythagoras developed a system of numbers that remans foundational for math today. Approaching mathematics with near-religious fervor, Pythagoras is credited with the theorem that the sum of the squares on the legs of a right triangle will match the square of the opposite side.

Human explorers have used the tool of mathematics to explore their world for millennia. An Egyptian papyrus almost four

---

[412] Richard P. Feynman, *QED: The Strange Theory of Light and Matter* (Princeton: Princeton University Press, 1985).

thousand years old depicted the calculations of volumes and areas, one goal of integral calculus.[413]Ancient Greek mathematicians used the method of exhaustion, a mathematical concept essential to calculus, to calculate areas and volumes. Mathematicians realized the full potential of calculus in the eighteenth century, when Sir Isaac Newton developed it as part of his investigations in physics and geometry.[414] Gottfried Leibniz, focusing on the tangent problem, believed that calculus was a God-given explanation of change.[415] Like the math of quantum electrodynamics, calculus is older than time itself.

Does the human mind invent math? Or do human minds discover math on an eternal landscape as old or older than time? Acknowledging that God is a mathematician, the astronomer James Jeans believed that the Almighty created every aspect of the universe in mathematical terms.[416] Centuries earlier, Galileo expressed the same opinion: "The book of nature is written in mathematical language."[417]

Many physicists agree that mathematics is timeless and eternal. Heinrich Hertz, the first to produce and detect radio waves in the laboratory, described his personal reaction to his own mathematics. "One cannot escape the feeling that these mathematical formulas have an independent existence of their own, and they are wiser than

413 Annette Imhausen, "Ancient Egyptian Mathematics: New Perspectives on Old Sources," *The Mathematical Intelligencer* 28 (2006): 19–27, https://doi.org/10.1007/BF02986998.
414 George F. Simmons, *Calculus Gems: Brief Lives and Memorable Mathematics* (Providence, RI: American Mathematical Society, 2007).
415 Maria Rosa Antognazza, *Leibniz: An Intellectual Biography* (Cambridge: Cambridge University Press, 2014).
416 James Jeans, *The Mysterious Universe* (Cambridge: Cambridge University Press, 2010).
417 Margaret L. Lial and Charles D. Miller, *Beginning Algebra* (New York: HarperCollins, 1992).

even their discoverers, and you get more out of them than was orig-
inally put into them."[418]

Einstein referred to the same phenomenon: "The eternal mystery of the world is its comprehensibility."[419]

**Despite being an "imaginary number," the square root of
minus one is a significant factor in quantum theory.**

We should repeat this vital observation. The square root of
minus one is an imaginary number, invented by mathematicians as
a random thought, produced by a human mind playing a mathematical game. But imaginary numbers play a key role in the mathematical description of quantum physics. What mind conceived of
this "complex number" and its role in quantum mechanics in the
first place?

Plato's Realm of Forms

Mathematicians who believe that all equations exist in a timeless
realm consider themselves to be Platonists.

[418] Davies, *The Mind of God*, 145.
[419] Albert Einstein, "Physics and Reality," *Journal of the Franklin Institute* 221, no. 3
(March 1936): 349–382, https://doi.org/10.1016/S0016-0032(36)91047-5.

Considered the leading Greek philosopher of the
classic period, Plato established an academy that
taught mathematics as well as philosophy.

Plato saw the physical world as fleeting and impermanent. But
he believed in an eternal world of ideal forms, a heaven-like domain
populated by eternal and unchanging ideas.[420] In Platonic theory,
mathematical concepts inhabit that higher dimension and shape
our physical world.

The Oxford mathematician Sir Roger Penrose is a Platonist.
He states, "It [the reality of mathematical concepts] is as though
human thought is…being guided toward some eternal external
truth."[421] As a leading example, Penrose points to the modern

---

[420] Penrose, *The Road to Reality*, 11.
[421] Roger Penrose, *The Emperor's New Mind* (Oxford: Oxford University Press, 2002).

prominence of "complex numbers," a number that includes an actual number combined with an "imaginary number."[422] The square root of minus one is a typical imaginary number.

***

Complex numbers sound like the mathematical equivalent of a unicorn. And yet, complex numbers simplify equations describing important aspects of electricity and quantum physics. Sir Penrose (knighted for his contribution to our understanding of black holes) would tell us that complex numbers are essential in understanding the real world: "[T]here often does appear to be some profound reality about these mathematical concepts, going quite beyond the mental deliberations of any particular mathematician."[423]

Penrose also believes that great works of art and music have "some kind of prior ethereal existence," as described in his 1989 book.[424] In an interview discussing if math was discovered or invented, Penrose also stated, "I think beauty is a clear guide to truth.... I am definitely sympathetic to all three of the Platonic ideals.... And then the moral, I would see even more so, probably."[425]

Visiting the Mathematical Library

Visits to the mathematical mindscape can produce dramatic and sudden moments of insight, unveiling previously unknown mathematical truths.

---

[422] Penrose, *The Road to Reality*, 13.
[423] Penrose, *The Emperor's New Mind*, 186.
[424] Penrose, *The Emperor's New Mind*, 97.
[425] "Roger Penrose: Mathematician," October 10, 2016, in *Why are we here?*, produced by Tern Television, https://www.whyarewehere.tv/people/roger-penrose/.

Paul Davies, in his classic *The Mind of God*, describes several breakthrough moments when solutions came to the mind abruptly.[426] As one mathematician expressed it, an answer to a riddle that had defied his understanding arrived "like a sudden flash of lightning."[427] Another mathematician described an epiphany as he boarded a bus. "At the moment when I put my foot on the step, the idea came to me, with nothing in my former thought seeming to have paved the way for it."[428] He was so confident that his brief insight had solved the problem, he put it in the back of his mind and completed the mathematical proof at his leisure.

Penrose recounts a similar incident during his efforts to understand black holes and space-time singularities (for which he won the Nobel Prize). He was about to cross a busy road when a crucial concept flashed into consciousness, but only for a second. It was only later that he became aware of a "curious feeling of elation, mentally recounted the events of the day, and remembered the brief inspirational flash."[429]

For just a few moments, his mind had made a visit to a timeless, eternal realm.

Timeless and Eternal

Hard-core Darwinists consider human intellectual ability to be the product of biological evolution. Why, we should ask, would the unfolding of our intelligence through evolution open the door to mathematical heaven? We might imagine that our brains have

---

[426] Davies, *The Mind of God*.
[427] Davies, *The Mind of God*.
[428] Davies, *The Mind of God*.
[429] PLEASE UPDATE CITATION. "Interview with Roger Penrose," The Nobel Prize, March 2021, https://www.nobelprize.org/prizes/physics/2020/penrose/169729-penrose-interview-march-2021/.

evolved in response to environmental pressures, enabling us to find food and mates and avoid predators. But our mathematical abilities are clearly overkill.

We can understand black holes millions of light-years away. We can identify the nature of quarks that make up the nucleus of the atom. The unreasonable efficacy of mathematics in the natural sciences has an obvious source: our Creator is the ultimate mathematician.

We Have Met an Additional Element of Human Consciousness

The brain
Neural networks
Electromagnetic fields
The connectome
Mathematics, the mind's eye

Supplement Twenty

Explore the relationship between math and science. Which one is more likely to lead to scientific breakthroughs—the isolated discovery of new equations, or observations and experimental results that demand new mathematical explanations?

CHAPTER TWENTY-ONE

Psi

Can the mind extend beyond the brain and connect with other minds?

Mind Beyond Brain

Evidence for the mind extending beyond the brain challenges the material worldview. Skeptical materialists see the universe as just matter in motion and the mind as just heat off the wires, the product of material neurons and the flow of neurochemicals. They missed the memo that matter is a derivative product of the big bang, dependent on the Higgs field and gluons of the strong force for its production.[430] Matter, in other words, is a secondary substance, hardly the ground of being. The divine mind created the Higgs mechanism and the gluons of the strong nuclear force that control the behavior of matter.

The term "psi" is derived from the twenty-third letter of the Greek alphabet. As a word, psi means "psyche" or "soul."[431]

[430]   Michael Riordan, *Hunting of the Quark: A True Story of Modern Physics* (New York: Simon and Schuster, 1987).
[431]   Russell Targ, *The Reality of ESP: A Physicist's Proof of Psychic Abilities* (Wheaton, IL: Quest Books, 2012), 11.

While serving in the Prussian army in 1892, Hans Berger's horse threw him into the oncoming path of a horse-drawn artillery transport, threatening him with immediate trauma and death. That same day, his sister dispatched a telegram; she'd had a premonition that Hans had been seriously injured. Convinced that a telepathic connection had occurred between himself and his sister, Hans spent his entire career investigating the possibility that the brain is an electromagnetic organ. In 1924, he produced the first EEG, placing electrodes on the scalp that documented electromagnetic waves coursing across the brain.[432]

Berger believed he had experienced a telepathic connection with his sister. In subsequent years, numerous scientists have attempted to prove that concept. Areas of active study by investigators worldwide include telepathy (transfer of thoughts or feelings between individuals by means other than the classical senses), precognition (perception of information about future places or events before they occur), and clairvoyance (obtaining information about places or events in remote locations, by means unknown to current science).

Material skeptics have derided the scientific study of the paranormal, but the evidence for "psi" is significant. For example, Dean Radin, chief scientist at the Institute of Noetic Science, points to the supporting evidence at the Princeton Engineering Anomalies Research laboratory, demonstrating that intense mental focus has been witnessed to change the distribution of zeroes and ones as produced by a quantum "random number generator."[433] The American physicist Russell Targ, as well as Radin, believe that the telepathic

---

[432]   Hans Berger, *Psyche* (Jena, Germany: Gustav Fischer, 1940), 6.
[433]   R. G. Jahn et al., "Correlation of Random Binary Sequences with Pre-Stated Operator Intention: A Review of a Twelve-Year Program," *Explore* 3, no. 3 (May 2007), 244–253, https://doi.org/10.1016/j.explore.2007.03.009.

transmission of information by a "sender" to a sensory deprived "receiver" has been demonstrated through scientific research.[434] Targ also summarizes the status of psi: "I believe it would be logically and empirically incoherent to deny the existence of some kind of human ability for direct awareness or experience of distant events that are blocked from ordinary perception, such experience being commonly known as ESP [extrasensory perception]."[435]

Question:
Do you believe the mind can extend beyond the brain, even "entangle" with the minds of others?

Psychic Phenomena

Investigators of the paranormal address a fascinating issue: can the human mind extend beyond the physical brain and exert mental influence at a distance? A British professor attending a meeting of the Science of Consciousness Conference in Tucson, Arizona, shared a personal story from his childhood during World War II. His mother woke him in the middle of the night, insisting that the family pack the car and drive immediately to her sister's home some two hours away. The mother explained, "The Japanese have just shot down my sister's husband (her brother-in-law) in an air battle. I want to be with her when the news arrives."[436]

They arrived at the sister's home late that night, and a telegram arrived in the morning. It was the official notice that the Japanese

---

[434] Dean Radin, *Entangled Minds: Extrasensory Experiences in a Quantum Reality* (New York: Paraview Pocket Books, 2006), 115.
[435] Targ, *Reality of ESP*.
[436] W. H. West, personal recollection, 2023.

had killed the husband in an air battle over Iwo Jima. Given this profound example of clairvoyance, the British professor acknowledged he had subsequently been a believer in psychic phenomena throughout his life. His question to the organizer of the conference was straightforward: "I know these phenomena are real. How do such things happen?"

He directed his question to Dean Radin. Radin is one of the leading voices in parapsychology, or psi research, and has documented his findings in several books, including *The Conscious Universe: The Scientific Truth of Psychic Phenomena, Entangled Minds, and Real Magic.*[437] He makes a persuasive case for the reality of such effects as psychokinesis (the ability to move physical objects with the mind), clairvoyance or remote viewing (where a person can gather information and receive imagery from a distant and unseen location), precognition (perceiving events or objects in the future), and telepathy (the ability to read or communicate with minds at a distance by extrasensory means), all of which refute the idea that the mind is confined to the brain. One of his favorite examples: mental focus can alter the behavior of a device called a random number generator.

Quantum random number generators rely on the intrinsic randomness of quantum mechanics to produce truly random numbers.[438] Random number generators play an important role in fields such as cryptography and secure communication. Repeated measurement of a qubit (quantum bit) containing a superposition of zeros and ones can produce these two numbers in equal proportions. In numerous well-executed studies, investigators have demonstrated that one person focusing intently on the process can

---

437 Dean Radin, *Real Magic: Ancient Wisdom, Modern Science, and a Guide to the Secret Power of the Universe* (New York: Harmony Books, 2018).
438 Radin, *Real Magic*, 154.

cause a significant shift in the relative frequency of the numbers. By looking intently for zeros, slightly more zeros appear—to a degree that meets the scientific definition of proof.[439] When a non-quantum computer program is used to produce the numbers, there is no effect, a strong hint that quantum mechanics plays an essential role in the ability of mind to extend beyond the brain.[440]

Telepathy Supported by Data

Researchers have also produced convincing data for mental telepathy. In the classic Ganzfeld experiment (*Ganzfeld* is a German term alluding to the "entire field" of ordinary perception), one individual focuses intently on an image in an effort to telepathically "send" information about the image to a "receiver."[441] That "receiver" has been "prepared" by spending thirty minutes in a state of sensory deprivation, ears covered by soundproof headphones emitting white noise and eyes covered by ping-pong balls delivering diffused red light to the retina. After half an hour of mental focus, investigators display four images and ask the receiver to identify the sender's image.

By chance, this experiment should produce a hit once in four tries, for a positive score of 25 percent. In the thirty years since the inception of Ganzfeld experiments, replicated in dozens of laboratories and thousands of sessions worldwide, the average success rate

439   Yang Liu et al., "Device-Independent Quantum Random-Number Generation," *Nature* 562 (September 2018): 548–551, https://doi.org/10.1038/s41586-018-0559-3.
440   Radin, *Real Magic*, 154.
441   Radin, *Real Magic*, 154.

has been consistently 33 percent. The magnitude might seem small, but by the statistical rules of science, it is profoundly significant.[442]

Mainstream science continues to resist these findings, and investigators continue to modify the experimental setup to overcome skepticism. One recently reported study is typical of the field.[443] Investigators at the University of Edinburgh attempted to optimize their study by selecting participants with self-reported creativity (such as musicians and artists) who also believed in psi. (Investigators classify believers as "sheep" and skeptics as "goats." Experiments with goats, who bring their material and skeptical worldview to the process, generate results that are typically negative.)

Target images and decoys were short video clips randomly selected from a pool of two hundred. Rather than depend on a human "sender" to view images and attempt to transmit them telepathically to the "receiver," their protocol eliminated both the sender and receiver and instead tested the ability to "see into the future," a concept known as precognition. The target was selected from a portfolio by a computer only after the participant had completed their sensory deprivation and reported their thoughts, feelings, and mental imagery in notes recorded for later review. To repeat for emphasis: to fully prevent bias, investigators completed the judging, recording, and uploading of results *before* a computer randomly selected the target image.

A computer then randomly presented the target clip and three decoy clips for the participant to review, who then scored

---

[442] L. Storm, P. E. Tressoldi, and L. DiRisio, "Meta-Analysis of Free-Response Studies, 1992–2008: Assessing the Noise Reduction Model in Parapsychology," *Psychological Bulletin* 136, no. 4 (2010): 471–485, https://psycnet.apa.org/doi/10.1037/a0019457.

[443] C. Watt et al., "Testing Precognition and Alterations of Consciousness with Selected Participants in the Ganzfeld," *Journal of Parapsychology* 84, no. 1 (2020), 21–37, http://doi.org/10.30891/jopar.2020.01.05.

each image for its relationship to their mental experience. Three experimenters conducted twenty trials each. The result: there were twenty-two direct hits in sixty trials (a 37 percent hit rate), well above the chance-based rate of 25 percent.[444]

Researchers have also studied precognition by measuring the electrical conductance of the skin before presenting computer images to a subject, some nonthreatening and others containing violent and disturbing content.[445] Skin conductance rose in anticipation of the disturbing images, while there was no change in skin conductance before a tranquil image. One subject interviewed on National Public Radio was convinced that the experiment had proven the reality of precognition, declaring, "I could see about three second into the future."[446]

In our discussion of neo-Darwinism, we have met Thomas Kuhn's classic book *The Structure of Scientific Revolutions*.[447] Scientists dedicate their careers to problems defined by the currently accepted worldview, described by Kuhn as a paradigm. The boundaries of a paradigm outline how the universe might behave and, equally important, what is beyond possibility.

Paradigms eventually fail to explain observed results, and inadequacies force the paradigm into crisis. Neo-Darwinism is facing a crisis as gene sequencing uncovers one unexpected finding after another. The materialist view of neuroscience is facing a crisis, as well. Max Planck, the founder of quantum mechanics, addressed the challenge posed by human consciousness in the British newspaper *The Observer* in January 1931: "I regard consciousness as fundamental. I regard matter as derivative from consciousness. We

---

[444] Watt et al., "Testing Precognition."
[445] Radin, *Real Magic*, 165.
[446] Radin, *Real Magic*, 165.
[447] Thomas S. Kuhn, *The Structure of Scientific Revolution*, 4th ed. (Chicago: University of Chicago Press, 2012).

cannot get behind consciousness. Everything that we talk about, everything that we regard as existing, postulates consciousness."[448]

Supplement Twenty-One

Altered memory resembles a form of "psi." Certain individuals have "highly superior autobiographical memory." Others have lost their memory altogether.

---

[448] J. W. N. Sullivan, "Interviews with Great Scientists," *The Observer*, January 25, 1931.

Mind Despite a Compromised Brain

The relationship of the mind with the material brain has been a topic of philosophic and scientific discussion since the days of the ancient Greeks. Is the mind the impermanent vapor of a hard-working brain? Or is the mind the reason the brain exists? Sometimes the mind functions in the face of a compromised or damaged brain. Does that provide a clue to the nature of their relationship?

The Brain-Damaged Mind

If the "mind is what the brain does"—the materialist view that our mental life emerges from the electrochemical activities of the brain like fog forming and dissipating on a creek or river valley—then the mind should totally depend on the integrity of the mind-producing brain. Is that always the case?

Under normal circumstances, a compromised brain correlates closely with abnormal mental function. When boxers deliver a knockout blow, the mental capacities of the vanquished opponent instantly go missing. With cardiac arrest and the loss of blood flow to the brain, loss of consciousness quickly follows. Individuals under the influence of alcohol and drugs commonly fail a sobriety test, indicating impaired cognition, while diseases of the brain such as Alzheimer's or progressive glioblastoma wreak havoc with mental function. But is the correlation between a challenged brain and a less-than-optimal mental function as tight as we might presume?

Three lines of evidence suggest that the mind can function despite a less-than-perfect brain. Children with hydrocephalus and skulls filled with cerebrospinal fluid can be left with minimal residual brain tissue. And yet, these children often display normal IQs.[449]

Patients ravaged by advanced dementia or rendered comatose by advanced cancer or other brain diseases can display remarkable recovery of mental function shortly before death.[450] This "terminal lucidity" is described by one observer as follows: "The (re-)emergence of normal or unusually enhanced mental abilities in dull, unconscious, or mentally ill patients shortly before death, including considerable elevation of mood and spiritual affectation, or the ability to speak in a previously unusual spiritualized and elated manner."[451] Is this evidence that an eternal mind is escaping the confines of a disordered brain?

Patients undergoing a near-death experience, with a near flatline EEG, can experience vivid, ordered thinking, sometimes involving encounters with the predeceased and detailed reviews of their life.[452] Sometimes these individuals return from their experi-

---

[449]   Roger Lewin, "Is Your Brain Really Necessary?" *Science* 210, no. 4475 (December 1980): 1232–1234, https://doi.org/10.1126/science.7434023.

[450]   Michael Nahm and Bruce Greyson, "Terminal Lucidity in Patients with Chronic Schizophrenia and Dementia: A Survey of the Literature," *Journal of Nervous and Mental Disease* 197, no. 12 (December 2009), 942–944, http://doi.org/10.1097/NMD.0b013e3181c22583.

[451]   Michael Nahm, "Terminal Lucidity in People with Mental Illness and Other Mental Disability: An Overview and Implications for Possible Explanatory Models," *Journal of Near-Death Studies* 28, no. 2 (Winter 2009): 87–106, https://digital.library.unt.edu/ark:/67531/metadc461761/.

[452]   Bruce Greyson, *After: A Doctor Explores What Near-Death Experiences Reveal about Life and Beyond* (New York: St. Martin Essentials, 2021).

ence with "veridical" information, such as visual observations or overheard conversations obtained when they were "out of body."

> **Question:**
> *If the mind is the emergent product of the matter of the brain, why does the mind seem to perform normally, or even "outperform," when the brain itself is far from normal?*

Hydrocephalus

Some unfortunate children suffer from hydrocephalus, otherwise known as "water on the brain," caused by an abnormal accumulation of cerebrospinal fluid around the brain. When excessive cerebrospinal fluid flows into the brain but can't escape, it causes the skull to expand. The pressure of the spinal fluid can compress the tissues of the brain to a small percentage of normal volume.[453] Fortunately, with early detection, surgeons can insert a shunt to drain the cerebrospinal properly, allowing these children to develop normally.

But if the surgery fails, the entire skull can fill with cerebrospinal fluid and compress the brain, creating severe degrees of neurological impairment.

In a 1970s study of six hundred cases of hydrocephalus, fluid had filled 95 percent of the cranial vault in sixty children.[454] Half of these children were severely impaired, but the remaining half had normal IQs. One student had a first-class honors degree in mathematics! In a subsequent report, French neurologists described

---

[453]   Kristopher T. Kahle et al., "Hydrocephalus in Children," *The Lancet* 387, no. 10020 (February 2016): 788–799, https://doi.org/10.1016/ S0140-6736(15)60694-8.
[454]   Lewin, "Is Your Brain Really Necessary?"

an individual with "massive ventricular enlargement" (ventricles connected to a cerebral aqueduct contain and regulate normal cerebrospinal flow). Despite massive compression of his cerebral cortex, this patient was married with two children and enjoyed a successful career as a civil servant.[455]

In even more dramatic cases, children as young as three years old undergo the surgical removal of their dominant cerebral hemisphere to relieve the symptoms of intractable seizures. A clinical account of one patient completing this surgery noted, "astonishingly, memory and personality develop normally." She suffered minimal impairment and has lived a normal life.[456]

A thoughtful essay by neurosurgeon Michael Engor describes an additional case about a patient named Katie: "I watched the CAT scan images appear on the screen, one by one. The baby's head was mostly empty. There were only thin slivers of brain—a bit of brain tissue at the base of the skull, and a thin rim around the edges. The rest was water.... [Katie] had little chance at a normal life."

He continues, "I cared for Katie as she grew up. At every stage of Katie's life so far, she has excelled. She sat and talked and walked earlier than her [twin] sister. She's made the honor roll. She will soon graduate high school."[457]

[455] Lionel Feuillet, Henry Dufour, and Jean Pelletier, "Brain of a White-Collar Worker," *The Lancet* 370, no. 9583 (July 2007): 262, https://doi.org/10.1016/S0140-6736(07)61127-1.
[456] Charles Choi, "Strange but True: When Half a Brain is Better than a Whole One," *Scientific American*, May 24, 2007, https://www.scientificamerican.com/article/strange-but-true-when-half-brain-better-than-whole/.
[457] Michael Egnor, "Science and the Soul," *Plough*, August 20, 2018, https://www.plough.com/en/topics/justice/reconciliation/science-and-the-soul.

These Observations Have Driven a Heated Debate

1. Is the brain enormously plastic, capable of rewiring despite major damage and distortion of cerebral tissue?
2. Pertinent to the topic of this chapter, do these cases provide evidence that the mind's dependence on the brain is relative rather than absolute?

Paradoxical or Terminal Lucidity

Patients with severe dementia can experience a remarkable remission in the hours or days prior to death.[458] A high percentage of caregivers in "memory units" have observed these remarkable events. Individuals who have been withdrawn and noncommunicative for protracted periods of time demonstrate surprising awareness of their environment, with normal behavior and speech. Remission of cognitive impairment has also been reported in patients with brain tumors, strokes, brain abscesses, and mental disorders, including advanced schizophrenia.

In 2020, researchers from three neurological institutes reported the results of an online questionnaire completed by caregivers between 2013 and 2019.[459] The survey was completed by 187 respondents, seventeen of whom provided detailed case reports. Lucidity was observed in 197 cases, including several who reported more than one case and seventeen who provided detailed case

---

[458] Alex Godfrey, "The Clouds Cleared: What Terminal Lucidity Teaches Us About Life, Death and Dementia," *The Guardian*, February 23, 2021, https://www.theguardian.com/society/2021/feb/23/the-clouds-cleared-what-terminal-lucidity-teaches-us-about-life-death-and-dementia.

[459] A. Batthyany and B. Greyson, "Spontaneous Remission of Dementia Before Death: Results from a Study on Paradoxical Lucidity," *Psychology of Consciousness: Theory, Research, and Practice* 8, no. 1 (2021): 1–8, https://psycnet.apa.org/doi/10.1037/cns0000259.

reports.. Almost two-thirds of the patients were unresponsive or unconscious most of the time. Ninety-three percent had marked difficulties with memory, attention, or focus.

The lucid episode was as short as ten minutes in twenty patients, ranged from ten to sixty minutes in forty-five, and lasted from several hours to several days in fifty-nine. Over 90 percent of patients died within seven days of their lucid period. More than 80 percent of patients appeared to experience a "full, albeit brief, reversal of profound cognitive impairment."

Marilyn Mendoza reported an example of terminal lucidity in *Psychology Today*. A ninety-one-year-old woman had been the victim of Alzheimer's disease for fifteen years and had shown no evidence that she recognized her daughter for the previous five years. Then, one day, "she started a normal conversation with her daughter. She talked about her fear of death [and] difficulties she had with the church and family members." She died a few hours later.[460]

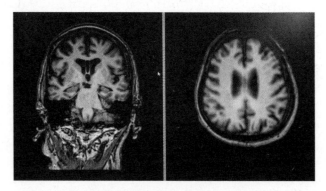

**This MRI of Alzheimer's disease reveals
extensive destruction of grey matter.**

---

460 Marilyn A. Mendoza, "Why Some People Rally for One Last Goodbye Before Death," *Psychology Today*, October 10, 2018, https://www.psychologytoday.com/intl/blog/understanding-grief/201810/why-some-people-rally-one-last-goodbye-death.

Given the advanced toll dementia has placed on healthcare systems, the National Institute on Aging convened a workshop in June 2018 to review evidence for and against paradoxical lucidity and to develop a research agenda to understand the mechanisms for this remarkable phenomenon.[461]

Some investigators consider paradoxical or terminal lucidity evidence of a resurgence in neurotransmitters or neural connections associated with approaching death. Others see it as a sign that minds transcend bodies and even the brain. As Alexander Batthyány, the Viktor Frankl Chair for Philosophy and Psychology at the International Academy of Philosophy in the Principality of Liechtenstein put it, when asked about Austrian neurology and psychologist Viktor Frankl's thoughts on the human person, "There's something indestructible and irreducible about human personhood." Was Frankl referring to a mindful soul?[462] Professor Frankl was a world-famous Viennese psychiatrist, philosopher, Holocaust survivor, founder of 'logotherapy and existential analysis.' He was also an author of an acclaimed bestseller, Man's Search for Meaning, considered by many as one of the most influential books ever written.

Deathbed visions and dreams are a common experience at the end of life.[463] Patients report visits with predeceased family members and beautiful images of the afterlife. Medical staff typically dismiss these experiences as hallucinations.

---

[461] Basil A Eldadah, Elena M. Fazio, and Kristina A. McLinden, "Lucidity in Dementia: A Perspective from the NIA," *Alzheimer's and Dementia* 15, no. 8 (August 2019):1104–1106, https://doi.org/10.1016/j.jalz.2019.06.3915.

[462] Zaron Burnett III, "Terminal Lucidity: The Researchers Attempting to Prove Your Mind Lives On Even After You Die, MEL Magazine, September 26, 2018, https://medium.com/mel-magazine/terminal-lucidity-the-researchers-attempting-to-prove-that-your-mind-lives-on-even-after-you-die-385ac1f93dca.

[463] Sue Brayne, Chris Farnham, and Peter Fenwick, "Deathbed Phenomena and Their Effect on a Palliative Care Team: A Pilot Study," *American Journal of Hospice and Palliative Medicine* 23, no. 1 (January/February 2006): 17–24, https://doi.org/10.1177/104990910602300104.

One dyed-in-the-wool materialist reported his personal experience with terminal lucidity. Grieving the loss of his mother who was in hospice in an "irretrievable coma," he was awakened from his sleep at her bedside "to find her reaching her hand out to me and she seemed very much aware. She was too weak to talk but her eyes communicated all…. Soon she closed her eyes again, this time for good. She died the next day." Despite the interaction, the man was still uncertain of the experience: "I really don't know how my mother managed those five minutes of perfect communion with me when, ostensibly, all of her cognitive functions were already lost. Was it her immortal soul? One last firestorm in her dying brain? Honestly, I'm just glad it happened."[464]

The time has come to review the concept that minds can escape the brain entirely, existing "out of body," even at the brink of death. In the next chapter, we will explore the rich literature of the near-death experience. The terminology may surprise you; the Apostle Paul was among the first to use it.

Supplement Twenty-Two

In this chapter, we have reviewed reports of patients with advanced dementia recovering normal brain function. In the chapter to follow, we will review the experience of individuals with no brain function at all undergoing a near-death experience. This supplement describes an even more remarkable anomaly: young children believe they have lived before, providing remarkable memories of their previous life. What explains the ability of young children to recall fact-based details of another life?

---

[464] Jesse Bering, "One Last Goodbye: The Strange Case of Terminal Lucidity," *Scientific American*, November 25, 2014, https://www.scientificamerican.com/blog/bering-in-mind/one-last-goodbye-the-strange-case-of-terminal-lucidity/.

CHAPTER TWENTY-THREE

The Near-Death Experience

Do near-death experiences provide a preview of eternal life?

Near-death experiences

Near-death experiences (NDEs) are anecdotal events that occur in a wide variety of circumstances.[465] Two aspects of these experiences argue for their validity. First, throughout history and in large numbers today, near-death survivors describe their brief view of heaven with identical language. They speak of otherworldly light, being out of the body, encountering the predeceased, undergoing a detailed life review, and even meeting the God of the universe.

Although only one in five survivors of an NDE recall a life review, those who do recall it provide a remarkably consistent account.[466] They see every event in their life instantly, as though those events are presented in a quantum superposition. More importantly, they experience their words and deeds from the perspective of second parties: did they love their fellow human beings or inflict injury or pain? How curious that unrelated individuals tell such a similar story. Few of us remember, or even know, the

---

[465]   Bruce Greyson, *After: A Doctor Explores What Near-Death Experiences Reveal about Life and Beyond* (New York: St. Martin Essentials, 2021).

[466]   Greyson, *After*, 40.

full impact of our words or deeds on our fellow human beings. Life reviews must draw on memories from a memory bank much greater than our own!

An auto mechanic suffered a terrible injury when his car jack failed, and the car crushed his chest. After weeks in the hospital at the brink of death, he returned to his home, where he obsessively continued to draw a letter *h*. At the insistence of his wife, he toured the library looking for the significance of *h* as a symbol. He returned with a biography of Max Planck and declared the solution to his obsession. An italic *h* stands for Planck's constant, a component of the equations of quantum mechanics. He declared that he had met Max Planck in the light.[467]

Tens of thousands of near-death stories provide similar reports. Students of the New Testament have heard these terms before. The Apostle Paul experienced an otherworldly light (per Luke in Acts) and an out-of-body experience ("whether in the body or out of the body, only God knows").[468] He declares we have both a physical and a spiritual body.[469] Paul asserts that love, even more than faith and charity, is the ultimate priority of life.[470] And he tells us more than once, as did Christ himself, to prepare for a life review.[471,472,473,474,475]

In future chapters, we will see that NDEs, deep meditation, and psychedelic-induced mystical encounters share a common

---

[467] Brandon R. Brown, *Planck: Driven by Vision, Broken by War* (Oxford: Oxford University Press, 2015).

[468] 2 Corinthians 112:7 (NIV).

[469] 1 Corinthians 15:44 (NIV).

[470] 1 Corinthians 13:13 (NIV).

[471] Romans 2:16 (NIV); Romans 14:10 (NIV); I Corinthians 4:5 (NIV); II Corinthians 5:10 (NIV); Matthew 12:36 (NIV).

[472] Romans 14:10 (NIV).

[473] 1 Corinthians 4:5 (NIV).

[474] 2 Corthians 5:10 (NIV).

[475] Matthew 12:36 (NIV).

foundation: they reveal that all is one, all is in God, and love is the currency of creation.

Views From the Outside

Since the time of Plato, individuals have reported a variety of inexplicable experiences associated with the process of death, challenging the materialist worldview that the mind is a short-term product of the brain.[476]

In the era of modern medicine, physicians routinely resuscitate patients approaching death. Patients surviving a NDE have registered their experience on the internet and described their NDE in well-respected books. Together, these reports give us a fascinating insight into what it might be like to make a brief trip to heaven.

Published NDEs

Mary Neal, an orthopedic surgeon, became wedged in her boat under a rock while kayaking in Chile, trapping her underwater for more than thirty minutes.[477] A long-standing Christian, she called for Jesus. She reported that Jesus held her in His arms and assured

---

[476] Plato, *Great Dialogues of Plato*, trans. W. H. D. Rouse (New York: Signet, 2015), 492.

[477] Mary C, Neal, *To Heaven and Back: A Doctor's Extraordinary Account of Her Death, Heaven, Angels, and Life Again* (New York: Random House, 2012).

her that she and her family would be okay. After multiple bone fractures and joint dislocations, the river released her body.

Immediately, her spirit left her body and a group of predeceased spirits celebrated her arrival in heaven. While enjoying this reunion, she could simultaneously hear her family begging for her to take a breath as they stood over her body on the bank of the river. She reluctantly returned to her body, took a breath, and then returned to her out-of-body reunion.

Finally, a member of her greeting party, perhaps a predeceased uncle, explained that it was not her time. After many surgeries and a prolonged healing process, she recovered to tell her story, both in interviews and a book entitled *To Heaven and Back*.

Butterfly Guide

Eben Alexander, a Harvard neurosurgeon developed a brain-throttling *E. coli* meningitis that left him near death for weeks.[478] He believed he might have had the deepest NDE in history, making multiple out-of-body trips to the center of heaven. His guide? A beautiful woman riding on a butterfly.

After his recovery, he reunited with his biological parents (he'd been adopted as a child), who informed him they had given up his baby sister for adoption as well. Sadly, his biological parents reported, his sister had died some years before. When they showed him a picture of his sister, he recognized her immediately—she was his butterfly-riding guide.

Dr. Alexander reported his experience in *Proof of Heaven: A Neurosurgeon's Journey into the Afterlife*.

---

[478] Eben Alexander, *Proof of Heaven: A Neurosurgeon's Journey into the Afterlife* (New York: Simon and Schuster, 2012).

Physician Reports

Many physicians have passed along NDE stories from their patients, and they have been reported widely in books and journals. One of their best-known reports includes a woman undergoing surgery for a brain aneurysm in an Atlanta hospital. Surgeons cooled her entire body and drained the blood from her brain. An anesthesiologist placed a sound-generating device in her ear to monitor her brain's response to sound and monitored her EEG, confirming the absence of brainwaves.[479]

Following her recovery, she described her surgery in stunning detail. She had observed her procedure from the ceiling, having a classic NDE. She recalled traveling through a tunnel toward a bright light, where she met her deceased grandmother and other predeceased relatives from both sides of her family.

In a similar case, a Hispanic woman suffered a cardiac arrest in a San Diego hospital. After her recovery, she reported an out-of-body experience with an unusual detail: while out of body, she had observed a blue tennis shoe on a window ledge of the hospital, scuffed, with a lace folded beneath. No staff member could see such a shoe from her room or the ground. A hospital employee went floor to floor, looking out each window of the hospital until she finally found a shoe precisely as described. There was only one way the patient could have seen the shoe: from an out-of-body perspective.[480]

A social worker in New York suffered a heart attack, and an ambulance rushed him to the nearest coronary care facility. He was

---

[479] Michael Sabom, *Light and Death: One Doctor's Fascinating Account of Near-Death Experiences* (Grand Rapids, MI: Zondervan, 1998), 37.

[480] Kenneth Ring and Madelaine Lawrence, "Further Evidence for Veridical Perception During Near-Death Experiences," *Journal of Near-Death Studies* 11 (1993): 223–229, https://doi.org/10.1007/BF01078240.

immediately shielded by drapes for placement of a stent, and the drapes blocked his view of the staff who entered from an opposite door. Looking up at the corner of the room, he described a lovely lady inviting him to join her on the ceiling.

In an instant, he was beside her, looking down on his procedure. Then, he described hearing an automated machine-like voice saying: "One-two-three, clear." Instantly, he was shocked and back in his body.

Following his recovery, having never met the physician who placed his stent, he accurately described him as a bald, "chunky fellow." How could he make this identification? He could see the top of the cardiologist's head from his perch up on the ceiling.[481]

A patient who had a cardiac arrest during a surgical procedure and temporarily lost his heartbeat later asked his surgeon why he was "flying" during the operation. The surgeon realized that the patient was accurately describing his movements during the procedure. To avoid contaminating his surgical gloves, he had given directions by pointing with his elbows. The surgeon concluded that the patient would only have made that observation if "out of body."[482]

A patient who had a heart attack outside a Denmark hospital arrived in the hospital ER cold and blue. After a difficult resuscitation, he spent weeks in ICU. As one nurse passed by, he recognized her from his resuscitation in the ER, declaring: "You placed my dentures in a drawer."[483]

---

[481] Sam Parnia et al., "AWARE-AWAreness During Resuscitation—A Prospective Study," *Resuscitation* 85, no. 12 (December 2014):1799–1805, https://doi. org/10.1016/j.resuscitation.2014.09.004.

[482] Greyson, *After*, 65.

[483] Pirn van Lommel et al., "Near-Death Experiences in Survivors of Cardiac Arrest: A Prospective Study in the Netherlands," *The Lancet* 358, no. 9298 (December 2001): 2039–2045, https://doi.org/10.1016/S0140-6736(01)07100-8.

Finally, a medical intern had to resuscitate a patient by himself because the other interns and residents assigned to the resuscitation team failed to show. He resuscitated his patient repeatedly until the patient finally stabilized.

Before discharge, the patient asked to meet with the intern to thank him for saving his life. Then he added a comment: "I hate you felt abandoned by your team." While out of body, he sensed the intern's emotions. He made another accurate observation: "Here I was dying in front of you.... And then you ate my lunch!"[484]

Common Features of an NDE

Raymond Moody was the first physician to collect an extensive series of near-death experiences in his classic book *Life after Life*.[485] Dr. Bruce Greyson has also made important contributions to the field. Individuals interested in the NDE consider his book *After: A Doctor Explores What Near-Death Experiences Reveal About Life and Beyond*[486] to be a classic summary of the field. Together, these experts have developed a scoring system to assess the depth of an NDE. Their system includes being out of body, seeing an other-worldly light, sensing that you are in a spiritual body, meeting the predeceased, encountering a god-like figure, and undergoing a life review. They have also described other common features that can include a sense of receiving special knowledge and firm instruction that to love fellow humans is the purpose of life.

Researchers have followed NDE survivors for decades and have reported consistent and substantial changes in their lives: First,

---

[484] Sam Parnia and Josh Young, *Erasing Death: The Science that is Rewriting the Boundaries Between Life and Death* (San Francisco: HarperOne, 226).
[485] Raymond Moody, *Life After Life* (San Francisco: HarperOne, 2001).
[486] Greyson, *After*.

they are reluctant to report their experience. Worried that people will consider them mentally unbalanced, they wait an average of fourteen years before cautiously speaking about their experience. Importantly, they also display more altruistic behavior, less fear of death, and often a sense of mission.[487]

Multiple aspects of near-death reports argue for the authenticity of NDEs, in contrast to dreams and other central nervous system phenomena such as hallucinations:[488]

1. NDEs are never interrupted but always reach a conclusion, typically ending in a statement that "it's not your time." Dreams rarely come to a logical conclusion.

2. Individuals report heightened thoughts and senses when the brain should be barely hanging on. There is evidence that the brain is still functioning after the pupils are fixed and dilated, but neuroscientists doubt that a brain at the point of death would produce a heightened sense of reality and a logical and consistent script.

3. NDEs happen under deep anesthesia—again, not a moment when the material brain should be hyperconscious.

4. When an NDE involves encounters with family members, those individuals are uniformly predeceased.

5. The deaf hear (telepathically).

6. Those with colorblindness see colors for the first time.

7. Individuals blind since birth see for the first time, one even describing the wedding ring she had never seen before.

8. Children younger than four tell comparable stories, even drawing pictures of the angels who accompanied them to heaven.

---

[487] van Lommel et al., "Near-Death Experiences."
[488] Greyson, *After*, 54.

During a near-death experience, time loses its relevance. Individuals feel as though they have received special knowledge and even a sense of mission. Thinking seems different from the normal Earthbound process. Forty-seven percent report clearer-than-normal thinking, 38 percent report faster thinking, and 29 percent believe their thinking during the NDE was more logical.[489]And all this happens while the brain is nearly dormant. Skeptics consider this a comfort program executed by a dying brain. That explanation, like all other proposed natural mechanisms for the NDE, seems to fall far short.

The NDE world is expanding. The terminology now includes "shared death experiences," reports that a loved one or caregiver experienced visions or events shared by the dying person, "spiritually transformative experiences," "NDE-like experiences," and "psilocybin-induced experiences."

Experts are beginning to suspect a link between "NDEs, mystical experiences, the visions of prophets and apostles in the Holy Scriptures, inspiration in poetry, art and music, meditation, hypnosis, neurological diseases (such as temporal lobe epilepsy), and psychiatric disorders."[490]

Shakespeare said it best in *Hamlet*: "There are more things in heaven and earth, Horatio, than are dreamt of in your philosophy."

Supplement Twenty-Three

The Apostle Paul's conversion on the road to Damascus is a widely known and beloved story, foundational to the Christian

---

[489] Greyson, *After*, 54.
[490] Shared Crossing Research Initiative, "Shared Death Experiences: A Little-Known Type of End-of-Life Phenomena Reported by Caregivers and Loved Ones," *American Journal of Hospice and Palliative Medicine* 38, no. 12 (April 2021):1479–1487, https://doi.org/10.1177/10499091211000045.

faith. Scattered through his letters are hints that he experienced the equivalent of an NDE (otherworldly light, new knowledge, divinely ordained mission, the priority of love, etcetera).

Compare the language of the Apostle Paul with near-death reports today. Voices across two millennia use similar, unrehearsed language to describe a near-death or spiritually transformative out-of-body experience. Does the concordance of ancient and modern voices lend credence to the reality of the near-death experience?

Spiritual Mind

Does the Holy Spirit gently woo and coach us? Is there a "Holy Spirit Field"?

Holy Spirit Field

The Maker of the Heavens and the Earth has repeatedly demonstrated His power.

1. His blueprints created a potentially infinite universe from something smaller than an atom.
2. They converted colorless clouds of hydrogen gas into the splendor of the nighttime sky.
3. His table of elements became the chemistry of life as DNA, proteins, and complex biochemical pathways appeared on the early Earth.
4. Gene and protein sequences produced neurons, synapses, and neurotransmitters to connect human minds with the mind of their Creator.

Would the God who inspired Holy Scripture, who opened His world to scientific study and gave us the gift of His son, then leave us neglected and alone? Or has He dispatched His spirit to promote

His moral code, to remind us of His love, and to provide ongoing inspiration and instruction?

Practitioners of deep meditation and patients receiving psychedelics tell us a divine mind waits for our connection. As Jesus promised and God has fulfilled, the mind of God is closer than we can imagine.

A mother's love is a vital fuel for the normal development of an infant's brain.[491] As infants soak up their world, hugs and kisses nurture their happiness and growth. Communist Romania consigned orphans to heartless children's homes, where no one responded to their tears. The devastating result: their brains remained underdeveloped. As adults, they had a lifetime of difficulties bonding with other humans.[492]

Adults never outgrow the need for love, and many of us find it in our involvement with religion. Teenagers involved in faith practice are less likely to use drugs.[493] People of faith enjoy longer and healthier lives.[494]

---

[491] Jeffry A. Simpson, W. Andrew Collins, and Jessica E. Salvatore, "The Impact of Early Interpersonal Experience on Adult Romantic Relationship Functioning: Recent Findings From the Minnesota Longitudinal Study of Risk and Adaptation," *Current Directions in Psychological Science* 20, no. 6 (December 2011): 355–359, https://doi.org/10.1177/0963721411418468.

[492] Melissa Fay Greene, "30 Years Ago, Romania Deprived Thousands of Babies of Human Contact," *The Atlantic*, July/August 2020, https://www.theatlantic.com/magazine/archive/2020/07/can-an-unloved-child-learn-to-love/612253/.

[493] F. F. Marsiglia et al., "God Forbid! Substance Use Among Religious and Nonreligious Youth," *American Journal of Orthopsychiatry* 75, no. 4 (2005): 585–598, https://doi.org/10.1037/0002-9432.75.4.585.

[494] Marino A. Bruce et al., "Church Attendance, Allostatic Load and Mortality in Middle Aged Adults," *PLOS One* 12, no. 5 (May 2017): article e0177618, https://doi.org/10.1371/journal.pone.0177618.

Involvement with religion has beneficial impact on the health and happiness of children. Teenage church-goers are less likely to experiment with drugs and more likely to volunteer in their community.

The Holy Spirit is as real as any force of physics. Without the stimulation of light waves, we would never develop the structures for sight. Without the stimulation of sound waves, we would never develop the physical infrastructure for hearing. Without the influence of the Holy Spirit, we would never know the fullness of love.

Is it unrealistic to propose another invisible field? Wave your hand through the air. Did you feel the presence of untold numbers of Higgs bosons? How about the nodes of space-time? They surround us in vast quantity, but we are never aware they exist. And yet we live on a material planet because those Higgs bosons delivered mass to matter. We stand with our feet on the floor because mass warps the geometry of space-time.

Love is no accidental product of evolution promoting the survival of the fittest. Love is the currency of creation, a gift of the Holy Spirit.

> **Question:**
> *Does the Holy Spirit inspire us in the presence of beauty and art and encourage us in times of difficulty? Would there be love without the continued stimulation of God's spirit in the world?*

Only Love Matters in the End

Mathematicians explore a mental mindscape and find the equations of creation. Tibetan monks practice deep meditation—exploring their own mindscape—and experience oneness with the universe. Patients treated with psilocybin, the ingredient in magic mushrooms, feel the presence of a higher power and oneness with the universe as well. Cancer patients participating in these experiences escape their anxiety and depression. They realize that "love [is] the only consideration,"[495] and they lose their fear of death. Individuals experiencing an NDE deliver the same message: love is the purpose of life.

There is a central theme to these observations: God did not intend life to be a one-way trip. Instead, He planned a round-trip

---

[495] Kelly Tatera, "Psychedelic Mushrooms Reduce Anxiety and Depression in Cancer Patients," The Science Explorer, December 15, 2015, https://www.thescienceexplorer.com/psychedelic-mushrooms-reduce-anxiety-and-depression-in-cancer-patients-573.

process. The divine mind created matter—and its most complex configuration, the human brain—to create creatures with a higher mind. God is spirit, the ultimate consciousness, and He created human beings in His image. We are self-aware and God-aware and instinctively understand that God is the source of love.

The Science of Love

What can science tell us about love? Is it a gift of a spiritual field, a phenomenon we might call the Holy Spirit?

In the early moments of life on planet Earth, the simplest of creatures began to respond to light. Vision is so central to life that God stored a regulatory complex for the eye, now known as *PAX6*, in the ancient genome. Regulatory control genes have supervised the construction of eyes in more than forty separate species,[496] beginning with the trilobite more than 540 million years ago.

When we look at an octopus, it looks back—using a camera eye like our own, even though the last common ancestor of the octopus and humans lived more than four hundred million years ago.[497] God wanted His creatures to see creation.

Eyesight is an amazingly complex process. A lens filled with the protein crystallin focuses light on the retina in the back of the eye. Optic nerves carry the visual information to the occipital lobes. More than forty separate neural networks slice and dice the input. One network analyzes vertical lines, another network recognizes horizontal lines, and while another responds to edges. Somehow, the brain reassembles the image and presents it to our

---

[496] Peter W. H. Holland, H. Anne F. Booth, and Elspeth A. Bruford, "Classification and Nomenclature of All Human Homeobox Genes," *BMC Biology* 5 (2007): article 47, https://doi.org/10.1186/1741-7007-5-47.

[497] Peter Godfrey-Smith, *Other Minds: The Octopus, The Sea, and the Deep Origins of Consciousness* (New York: Farrar, Strauss and Giroux. 2016), 13.

conscious mental field. An elaborate material system supports our ability to see.[498]

Would we have a visual system if there was no light? Mexican tetra, fish living in dark caves, provide one answer to that question. Their DNA has inactivated the genes previously used for vision. Since they have no stimulation by light, there is no reason to maintain a visual apparatus.[499]

The Iberian mole living in the dark underground has also abandoned its sight, covering its eyes with skin.[500] Living in ocean waters, cetaceans (whales and dolphins) make limited use of their conventional eyesight. Instead, they have developed some of the most complex and unique acoustic capabilities on Earth—piercing the darkness with sound.[501]

Hearing is another vital sense shared throughout the animal kingdom. The ability to hear requires another elaborate neural infrastructure, as sound waves reaching the eardrum trigger oscillations in hair cells.[502] Those oscillations send signals to the temporal lobe of the brain. Like vision, multiple neural networks analyze those sound waves before sending a coherent sound to the conscious mental field. Sound waves enrich life with the cries and howls of untold animal species and the medium of human speech.

---

[498] "How the Eyes Work," National Eye Institute, last updated April 20, 2022, https://www.nei.nih.gov/learn-about-eye-health/healthy-vision/how-eyes-work.

[499] Thomas E. Dowling, David P. Martasian, and William Jeffery, "Evidence for Multiple Genetic Forms with Similar Eyeless Phenotypes in the Blind Cavefish, *Astyanax Mexicanus*," *Molecular Biology and Evolution* 19, no. 4 (April 2002): 446–455, https://doi.org/10.1093/oxfordjournals.molbev.a004100.

[500] F. David Carmona et al., "Retinal Development and Function in a 'Blind' Mole," *Proceedings of the Royal Society B* 277 no. 1687 (December 2009): 1513–1522, https://doi.org/10.1098/rspb.2009.1744.

[501] Yong-Yi Shen et al., "Parallel Evolution of Auditory Genes for Echolocation in Bats and Toothed Whales," *PLOS Genetics* 8, no. 6 (June 2012): e1002788, https://doi.org/10.1371/journal.pgen.1002788.

[502] Ellen A. Lumpkin, Kara L. Marshall, and Aislyn M. Nelson, "The Cell Biology of Touch," *Journal of Cell Biology* 191, no. 2 (October 2010): 237–248, https://doi.org/10.1083/jcb.201006074.

It would seem a fair assumption to believe that the ability to hear would not exist without the stimulation of sound waves.

Love, as much as sight and sound, is a necessity for a happy human life. Everyone needs to love and be loved. As Thomas Aquinas, and Aristotle before him, defined it, to love is to will the good of another.[503] The philosopher Gottfried Leibniz said the same: love is "to be delighted by the happiness of another."[504] C. S. Lewis analyzed love in depth in his classic book on the subject.[505]

What We Learned About Love from Romanian Orphans

As with sight and sound, there is a material, biological foundation for love. The limbic system, a group of interconnected structures deep within the brain, controls behavioral and emotional responses. The amygdala binds emotions with memory and attention. Neurochemicals support love. Dopamine activates the brain's reward circuit, while oxytocin, the "bonding hormone," strengthens feelings of attachment.[506]

The normal development of the infant is more dependent on love than we can imagine. Between 1965 and 1989, the Romanian communist Nicolae Ceaușescu trapped an estimated one hundred thousand Romanian children in orphanages called "children's

503 Tom Neal, "To Will the Good of the Other," World on Fire (blog), February 24, 2016, https://www.wordonfire.org/articles/to-will-the-good-of-the-other/.

504 Maria Rosa Antognazza, *Leibniz: An Intellectual Biography* (Cambridge: Cambridge University Press, 2014).

505 C. S. Lewis, *The Four Loves* (New York: Harcourt, 1960).

506 D. D. Francis et al., "Naturally Occurring Differences in Maternal Care are Associated with the Expression of Oxytocin and Vasopressin (V1a) Receptors: Gender Differences," *Journal of Neuroendocrinology* 14, no. 5 (May 2002): 349–53, https://doi.org/10.1046/j.0007-1331.2002.00776.x.

houses."[507] Caretakers failed to give those children any loving attention. When diapers needed changing, the staff applied them in an assembly-line fashion. When children cried, no one responded to their tears.

When the condition of these children became known to the outside world following the fall of the Romanian Communist Party in 1989, the outrage was deafening.

Communist Romania housed orphans in government homes and deprived them of any attention resembling a mother's love. Following the overthrow of the government, MRI scans of these orphans revealed under-developed brains. A high percentage of these children had difficulty bonding with other humans as they entered adulthood.

---

[507] Shaun Walker, "Thirty Years On, Will the Guilty Pay for Horror of Ceaușescu Orphanages?" *The Guardian*, December 15, 2019, https://www.theguardian.com/world/2019/dec/15/romania-orphanage-child-abusers-may-face-justice-30-years-on.

Despite adequate nutrition, these Romanian orphans were both short and underweight. MRI studies of their brains revealed a shortage of grey matter—they had underdeveloped brains. EEG tracings confirmed the worst fears: they had profoundly compromised brain activity.[508] Even more tragic, despite adoption by loving families, many of these orphans had lifelong difficulty bonding with another human.[509]

Long-Term Effects from a Lack of Love

Depriving an infant of its mother's love has created heartbreaking effects in another primate. Scientists separated baby rhesus monkeys from their mothers, substituting maternal facsimiles made of wire and wood. These motherless primate infants swayed and twirled in their cages and self-mutilated—demonstrating their need for a mother's love as well.[510]

When parents shower an infant with love and provide a consistent parental bond, their neural pathways thrive. As one observer described it, their brain connections "multiply, intersect, and loop through remote regions of the brain like a national highway system under construction."[511] But in the brain of a neglected baby—a baby lying alone, staring at a ceiling, no one responding to its wet diaper, feeling unwanted week after week, year after year—fewer

508   Eliot Marshall, "Childhood Neglect Erodes the Brain," *Science*, January 26, 2015, https://www.science.org/content/article/childhood-neglect-erodes-brain. https://www.science.org/content/article/childhood-neglect-erodes-brain, January 26, 2015. Scan findings

509   Greene, "30 Years Ago,"

510   Harry F. Harlow, Robert O. Dodsworth, and Margaret K. Harlow, "Total Social Isolation in Monkeys," *PNAS* 54, no. 1 (1965): 90–97, https://doi.org/10.1073/pnas.54.1.90.

511   Greene, "30 Years Ago,"

connections get built. Failing to provide a reliable source of affection and stimulation blunts the brain's development.

The absence of love has long-term mental and emotional consequences. One psychologist explained, "Parental love is a bit like oxygen. It's easy to take for granted until you see someone who isn't getting enough."[512]

There is a reason that television commercials for St. Jude Children's Research Hospital and the Shriners Hospital System are so effective. Images of young children defiantly happy despite severe disabilities or potentially fatal cancer remind us of the protection we may have felt in the presence of our mothers. Whether a child has cancer or a congenital disability, lame limbs, or cleft lip, we have an instinctive response to share our love. Obviously, family and staff shower the children at St. Jude and Shriner hospitals with love—but many donors choose to add additional love of their own.[513]

The Benefits of Divine Love

As children grow less dependent on their mother's love, they benefit from another, external source of emotional support provided by their involvement with religion. Adolescence is a critical period of cognitive and behavioral development. A highly regarded scientific study found that adolescents who attended religious services regularly were 12 percent less likely to have depressive symptoms and 33 percent less likely to use illicit drugs.[514]

A religious upbringing contributed to other positive outcomes: greater happiness, more volunteering in the community, and a

---

[512] Greene, "30 Years Ago,"
[513] Walter T. Hughes, *On Hallowed Ground: St. Jude Children's Research Hospital* (self-pub., Outskirts Press, 2018).
[514] Marsiglia et al., "God Forbid!"

greater sense of mission and purpose. Those who attended religious services were 87 percent more likely to have elevated levels of forgiveness. Those who prayed or meditated frequently were 38 percent more likely to volunteer in their community.[515]

The belief in a loving God—in a creator and a spiritual guide—provides vital support for children in a potentially hostile and pessimistic world. One of the motivating factors for authoring this book is the following statistic: 44 percent of eighteen-to-thirty-year-old Americans now report their religious affiliation as "none."[516] That could have ominous implications for the long-term physical and mental health of an entire generation.

A recent study by a Vanderbilt University professor followed the health and well-being of five thousand adults between the ages of forty and sixty-five and reported that people who attend worship services reduce their risk of premature death by 55 percent.[517] By failing to attend church in the preceding year, adults doubled their risk of premature death.

Involvement in a Religious Community Promotes Health and Longevity

Other studies of regular church attendance reported better sleep, reduced risk of depression and suicide, more stable marriages,

---

[515] Crystal Amiel M. Estrada et al., "Religious Education Can Contribute to Adolescent Mental Health in School Settings," *International Journal of Mental Health Systems* 13 (2019): article 28, https://doi.org/10.1186/s13033-019-0286-7.

[516] Timothy Beal, "Can Religion Still Speak to Younger Americans?" *Wall Street Journal*, November 14, 2019, https://www.wsj.com/articles/can-religion-still-speak-to-younger-americans-11573747161.

[517] Marino A. Bruce et al., "Church Attendance, Allostatic Load and Mortality in Middle Aged Adults," *PLOS One* 12, no. 5 (May 2017): article e0177618, https://doi.org/10.1371/journal.pone.0177618.

lower blood pressure, and longer life.[518] Belonging to a community, developing a stronger sense of compassion or empathy for others, and developing a feeling of holiness and being part of something greater than oneself had measurable health benefits. The human need for love hardly disappears with advancing age. Adults need loving family and friends. They also long for something greater than themselves—a force that helps them define their moral sense and their values.

Are Homo Sapiens Wired to Seek the Transcendent?

Unique among animals, human beings are inclined to listen for God and seek out mystical experiences. One observer proposed that a specific gene underlies this behavior. In *The God Gene: How Faith is Hardwired into Our Genes*, Dean Hamer, at the time director of gene structure and regulation at the National Cancer Institute in Bethesda, attributed our spiritual inclination to a single gene that regulates the level of neurochemicals in the brain (a stretch, given that half our genes are expressed in the brain).[519] Religious leaders challenged this proposal, suggesting that divine transformation underlies our inclination to religious belief.

Experience with psilocybin described in the following chapter undermines both positions. A single treatment with the purified ingredient in magic mushrooms convinces a majority of atheists to

---

[518] Kaytura Feliz Aaron, David Levine, and Helen R. Burstin, "African American Church Participation and Health Care Practices," *Journal of General Internal Medicine* 18 (November 2003): 908–913, https://doi. org/10.1046/j.1525-1497.2003.20936.x.

[519] Dean H. Hamer, *The God Gene: How Faith is Hardwired into Our Genes* (New York: Anchor Books, 2005).

abandon their atheist worldview.[520] Our religious proclivity seems hardly limited to a single gene or the intensity of our faith-based exposure. God is spirit, and He created us to be spiritual creatures.

Supplement Twenty-Four

Religious affiliation is declining in the US. Does religion still matter? The Asbury awakening, which involved Gen Z students, has drawn worldwide attention. The Methodist revival in eighteenth century England reversed a national decline in the rule of law and adherence of faith.

---

[520] Roland R. Griffiths et al., "Survey of Subjective 'God Encounter Experiences': Comparisons Among Naturally Occurring Experiences and Those Occasioned by the Classic Psychedelics Psilocybin, LSD, Ayahuasca, or DMT," *PLoS One* 14, no. 4 (2019): 1–26, https://doi.org/10.1371/journal.pone.0214377.

CHAPTER TWENTY-FIVE

Cosmic Consciousness

Near-death experiences suggest life survives death of the body. The effects of treatment with psilocybin and other psychedelics lend credence to this possibility. Can human beings connect with the mind of God as we open our minds to a mind at large, a cosmic consciousness where all is one, and love is the currency of creation?

Cosmic Consciousness

Every great religion embraces meditation as a pathway to truth and enlightenment.[521] Eastern religions use meditation to escape the sorrows and passions of the ordinary life. Their goal is to achieve nirvana, a state of perfect quietude, freedom, and the highest happiness, as well as the liberation from attachment to material goods and worldly suffering.[522]

Deep meditation can produce a mystical experience, an apex of religious experience worldwide. Sufi poets, Hebrew prophets, and medieval priests—not to mention adherents of ancient Eastern

---

[521] Madhav Goyal et al., "Meditation Programs for Psychological Stress and Well-being: A Systematic Review and Meta-Analysis," *JAMA Internal Medicine* 174 no. 3 (2014): 357–368, https://doi.org/10.1001/jamainternmed.2013.13018.

[522] Thich Nhat Hanh, *The Heart of the Buddha's Teaching: Transforming Suffering into Peace, Joy, and Liberation* (Berkeley, CA: Parallax Press, 1999).

traditions—have experienced the feelings of profound mystery, unity with all creation, and ecstasy that characterize mystical encounters. But while the traditions generated by different cultures have divergent practices and goals, mystical experiences demonstrate striking similarities across times and cultures.[523]

Meditation can take years to master. Now modern science is providing a shortcut. Psychotropic drugs can replicate the power of meditation and create a mystical experience in a single dose.[524] The ingredient in magic mushrooms, psilocybin, is gradually becoming a mainstream tool for personal discovery and psychiatric treatment.[525]

Neurologists have scanned the brains of meditating monks and volunteers treated with psilocybin, with consistent and profound results. Mystical experiences correlate with reduced blood flow to structures in the brain known as the "default mode network" that filter and confine us.[526]

By modifying the narcissistic focus on egotistical needs, meditation and psychoactive drugs open the mind to a greater reality: the unity of creation and the power of love.

---

[523] Frederick S. Barrett and Roland R. Griffiths, "Classic Hallucinogens and Mystical Experiences: Phenomenology and Neural Correlates," in *Behavioral Neurobiology of Psychedelic Drugs*, ed. Adam L. Halberstadt, Franz X. Vollenweider, and David E. Nichols (Berlin: Springer, 2018), 393–430, https://doi.org/10.1007/7854_2017_474.

[524] Michael Pollan, *How to Change Your Mind: What the New Science of Psychedelics Teaches Us About Consciousness, Dying, Addiction, Depression, and Transcendence* (New York: Penguin Books, 2018).

[525] Tanya Lewis, "Johns Hopkins Scientists Give Psychedelics the Serious Treatment," *Scientific American*, January 16, 2020, https://www.scientificamerican.com/article/johns-hopkins-scientists-give-psychedelics-the-serious-treatment/.

[526] Andreas Horn et al., "The Structural-Functional Connectome and the Default Mode Network of the Human Brain," *NeuroImage* 102, pt. 1 (November 2014): 142–151, https://doi.org/10.1016/j.neuroimage.2013.09.069.

Following a mystical experience, cancer patients lose their fear of death. Smokers stop smoking. Former atheists sense the reality of a higher power and drop the atheist label.[527]

> **Question:**
> *Our brain has a filter focused on the here and now, leading to an excessive obsession with self. Can we escape that filter and appreciate a greater reality—that we are a valued component of a cosmic-wide consciousness reflecting the reality of God?*

The Mystical Experience

In our review of the field of psi, we met investigators and protocols striving to demonstrate that the mind can extend beyond the brain, connecting with other minds (telepathy) and even seeing events that lie in the future (precognition). Psi suggests that our minds can unconsciously connect with other minds in an entangled universe, escaping the limitations of space and time.

Meditation and psychedelics resemble turbocharged psi, throwing open a door to an entangled universe and allowing individuals to experience the reality of the "mind at large," a cosmic consciousness many interpret to be God.

Eastern religions have long supported meditation as a route to a greater reality and a cosmic connection. In Tibetan meditation, the goal is to quiet the constant chatter of the conscious mind and lose oneself in a deeper, simpler reality. As one drifts into that inner

---

[527] Roland R. Griffiths et al., "Survey of Subjective 'God Encounter Experiences': Comparisons Among Naturally Occurring Experiences and Those Occasioned by the Classic Psychedelics Psilocybin, LSD, Ayahuasca, or DMT," *PLoS One* 14, no. 4 (2019): 1–26, https://doi.org/10.1371/journal.pone.0214377.

spiritual state—as the material world recedes like a fading dream—meditation progresses toward what practitioners call a spiritual peak.[528] Worries, fears, desires—preoccupations of the conscious mind—melt away. When this deeper consciousness emerges, the individual suddenly understands that the inner self is not an isolated entity. Their mind connects with all of creation. There is a sense of timelessness and infinity. They consider this inner self the very essence of being, what some call their soul, and they recognize they are part of everyone and everything in existence.[529]

The mystical experience can happen spontaneously without drugs or meditation,[530] but students of meditation hope to achieve mystical experiences on a regular basis. Those experiences are characterized by the following features:[531]

1. The boundaries of individual consciousness and identity (ego) diminish, leaving the individual with a boundless and infinite union with all that exists. The experience is referred to as "conscious unity."

2. With no definable identity or spatial recognition, time feels infinite, a stream of eternal moments. The limitations imposed by classical space and time disappear.

3. As personal identity disappears along with a sense of time and place, the remaining consciousness takes on aspects of a much more intricate and profound reality. Everything seems innately perfect and connected.

---

528 Tenzin Wangyal Rinpoche, *Awakening the Luminous Mind: Tibetan Meditation for Inner Peace and Joy* (New York: Penguin Random House, 2015).
529 Matthew Soclolov, *Practicing Mindfulness* (New York: Althea Press, 2018).
530 Ralph W. Hood Jr. "The Construction and Preliminary Validation of a Measure of Reported Mystical Experience," *Journal for the Scientific Study of Religion* 14, no. 1 (March 1975): 29-41, https://doi.org/10.2307/1384454.
531 William Richards, *Sacred Knowledge: Psychedelics and Religious Experiences* (New York: Columbia University Press, 2015).

4. Despite the vastness of existence, the individual feels like a valued component of the cosmic order. The result is feelings of ecstasy accompanied by an immense sense of gratitude and awe and a new sense of respect for the sacredness of life.

For most participants, these experiences are ineffable, impossible to describe with words. Although mystical experiences are always transient, following the experience, the participant perceives and interacts with the world in a new and often more balanced way. Having encountered God as the controlling power of the universe, they no longer fear death. They see death as a door into a greater reality filled with unconditional love and spiritual joy.

The Use of Psychedelics

It may take years of practice to achieve a mystical experience through meditation. Psychedelic drugs can achieve a mystical state in a matter of hours.[532] Medicalized mysticism, a rapidly growing field of psychiatry, took root in a double-blind, placebo-controlled trial at Harvard in 1962 involving twenty theology students. Half of them received the magic mushroom ingredient psilocybin, while the other half received a placebo. After taking their medicine and attending Good Friday mass, nine of the ten who had received the active agent believed they experienced a genuine spiritual encounter. Only one of the thirteen receiving the placebo described a similar experience.[533] Investigators have repeated this experiment many times with diverse groups at various locations, while also

---

[532] James Stephen, "The Harvard Psilocybin Project: A Retrospective," Truffle Report, September 25, 2021, https://web.archive.org/web/20211028115701/https://www.truffle.report/the-harvard-psilocybin-project-a-retrospective/.

[533] Pollan, *How to Change Your Mind*, 45.

keeping track of the Good Friday participants for years. Those treated with psilocybin still consider their original experience to have been genuinely mystical, making an important contribution to the development of their spiritual lives.

Psyche comes from the Greek word for "the mind." *Deloo* is a Greek verb meaning "to manifest." Psychedelics thus open or "manifest the mind."[534] The most effective psychedelics bind to the 5-HT2A serotonin receptor, ordinarily dedicated to serotonin. Lysergic acid diethylamide, usually called LSD, was synthesized in a lab. Psilocybin and ayahuasca are natural products extracted from "magic mushrooms" and the leaves of the *Psychotria viridis* plant containing N,N-Dimethyltryptamine, commonly known as DMT. Native cultures have used magic mushrooms and ayahuasca for thousands of years.

In the 1960s, Timothy Leary advised young people to "turn on, tune in, and drop out."[535] Authorities involved in the Vietnam War were not amused. The Harvard administration dismissed staff members, including psychologist and writer Leary, for generating controversy, not fulfilling their teaching obligations, and giving psychedelics to undergraduates. The government soon enacted regulatory hurdles to prevent further study with psychedelics.[536]

Fast-forward to the Center for Psychedelic and Consciousness Research at Johns Hopkins University, where psychiatrists have cautiously reactivated studies of psilocybin. The basis for their interest: psilocybin produces changes in thoughts, perception, and emotions, often triggering profound alterations of perceptions of

---

[534] Pollan, *How to Change Your Mind*, 18.
[535] Don Lattin, *The Harvard Psychedelic Club: How Timothy Leary, Ram Dass, Huston Smith, and Andrew Weil Killed the Fifties and Ushered in a New Age for America* (New York: Harper Collins, 2011).
[536] Lattin, *The Harvard Psychedelic Club*, 1504.

reality, reminiscent of naturally occurring mystical and spiritual experiences.

Since September 2020, the center has administered psilocybin to hundreds of volunteers in more than seven hundred sessions in rigorous, double-blind studies. The take-home results for normal volunteers have been unanimous: "Psilocybin produced large increases on self-rated questionnaires designed to measure naturally occurring mystical-type and insightful-type experiences."[537] Psychedelic treatment left participants more open to mystical experiences in the future without a requirement for further medication.

Volunteers report a sense of unity and interconnectedness with all people as "one pure consciousness," accompanied by a sense of "sacredness or reverence or preciousness for the experience."[538] There is a sense of encountering ultimate reality with a feeling that these experiences are more real and more true than everyday waking consciousness. Volunteers also describe a deeply felt positive mood, including the reality of universal love, joy, and peace as past and future collapse into the present.

In the words of one volunteer: "In my mind's eye, I felt myself instinctively taking on a posture of prayer in my head. I was on my knees, hands clasped in front of me, and I bowed to this force. I wasn't scared or threatened in any way. It was more about reverence. I was showing my respect. I was humbled and honored to be in this presence. This presence was a feeling, not something I saw or

---

[537] "Video: Psilocybin and Mystical Experience: Implications for Healthy Psychological Functioning, Spirituality, and Religion," Center for the Study of World Religions, September 29, 2020, https://cswr.hds.harvard.edu/news/2020/09/29/video-psilocybin-and-mystical-experience-implications-healthy-psychological.

[538] "Video: Psilocybin and Mystical Experience," Center for the Study of World Religions.

heard. I only felt it, but it felt more real than any reality I've ever experienced."[539]

Following such experiences, more than 60 percent of participants who considered themselves atheists felt the presence of a higher power and abandoned the atheist label.[540]

Researchers at NYU have given psilocybin to patients suffering from anxiety, depression, and "existential distress" at the prospect of dying from life-threatening cancer.[541] In rigorous placebo-controlled trials, eighty volunteers embarked on a psychic journey that, in many cases, brought them face-to-face with their cancer and their fears. Eight out of ten patients receiving the active ingredient, given to each of them in equal doses, showed clinically significant reductions in anxiety and depression, with effects enduring for at least six months.

The degree to which symptoms improved in these trials correlated with the intensity of the "mystical experience." Neuroradiologists have also identified a correlation with reduction in activity of the "default mode network," also known as "the ego network."[542] Both meditation and the ingredient in magic mushrooms seem to shut down that egocentric component of mental function. With that out of commission, the mind bypasses barriers and encounters a greater reality.

---

[539] Barrett and Griffiths, "Classic Hallucinogens," 397.

[540] R. R. Griffiths et al., "Psilocybin-Occasioned Mystical-Type Experience in Combination with Meditation and Other Spiritual Practices Produces Enduring Positive Changes in Psychological Functioning and in Trait Measures of Prosocial Attitudes and Behaviors," *Journal of Psychopharmacology* 32, no. 1 (2018): 49–69, https://doi.org/10.1177/0269881117731279.

[541] Stephen Ross et al., "Rapid and Sustained Symptom Reduction Following Psilocybin Treatment for Anxiety and Depression in Patients with Life-Threatening Cancer: A Randomized Controlled Trial," *Journal of Psychopharmacology* 30, no. 12 (2016): 1165–1180, https://doi.org/10.1177/0269881116675512.

[542] Horn et al., "The Structural-Functional Connectome."

One person reported his experience as follows: "God exists, that He is love, that He is calling us all to unity with Him." The writer describing the effects on their friend described the experience as "a glimpse behind the veil."[543] Others point to a new understanding that love is the most potent force on the planet.

Key Definition: The Default Mode Network

A default mode (ego) network forces the mind to focus on self, perhaps to promote our survival during physical or emotional stress. Through deep meditation and treatment with psychedelics, this egocentric neural network can be bypassed, opening the mind to a cosmic consciousness and a greater reality: all is one, and love is the currency of creation.

Editors devoted an entire edition of *The Journal of Psychopharmacology*, a prominent psychiatric journal, to these psilocybin studies.[544] Psychiatrists believe that psilocybin will become a beneficial treatment for a variety of mental disorders, including anxiety, depression, and addiction. In a single guided psychedelic session, psilocybin can alleviate depression more effectively than months of treatment with a conventional antidepressant.[545]

Studies also suggest that psilocybin can help alcoholics and smokers break their drug dependence. A Johns Hopkins study of smoking found that 80 percent of the volunteers had successfully

---

[543] Rod Dreher, "The Psychedelic Dante," The American Conservative, February 5, 2015, https://www.theamericanconservative.com/the-psychedelic-dante/.

[544] *Journal of Psychopharmacology* 30, no. 12 (December 2016).

[545] Guy M. Goodwin et al., "Single-Dose Psilocybin for a Treatment-Resistant Episode of Major Depression," *New England Journal of Medicine* 387, no. 18 (2022):1637–1648, https://doi.org/10.1056/nejmoa2206443.

quit tobacco use after six months, outshining the standard treatment using nicotine replacements.[546]

Learning that the exalted mystical experience of oneness (often called "absolute unitary being") reflects the reduction of activity in a specific region of the brain might seem to suggest that the benefits of meditation and psychedelics reflect a rewiring of the material brain. But materialists should delay their celebration. During mystical experiences, individuals universally report that what they experience seems many times more real than everyday reality. They believe they can perceive God, or pure consciousness, as the ultimate reality.[547]

As Aldous Huxley expressed it following a psychedelic treatment in 1954: "That interfering neurotic who, in waking hours, tries to run the show, was blessedly out of the way."[548] Huxley was referring to the phenomenon known as drug-induced ego dissolution, a family of acute effects produced by high doses of psychedelic drugs, typically reported as a loss of one's sense of self and self-world boundary. Huxley was a believer in the "mind at large," a universal, interactive form of consciousness extending well beyond any single brain. Others have called it "cosmic consciousness," the "Oversoul," or "Universal Mind," something that exists outside our brain—a property of the universe, like electrons and quarks or gravity.[549]

Do we have any lingering doubt about the human mind's ability to extend beyond the brain, even connect with oneness and the

[546] Tehseen Noorani et al., "Psychedelic Therapy for Smoking Cessation: Qualitative Analysis of Participant Accounts," *Journal of Psychopharmacology* 32, no. 7 (2018): 756–769, https://doi.org/10.1177/0269881118780612.

[547] Pollan, *How to Change Your Mind*, 71.

[548] Aldous Huxley, *The Doors of Perception and Heaven and Hell* (New York: Harper Perennial Modern Classic, 2009),

[549] Huxley, *The Doors of Perception*.

281

reality of God? Focused minds project images to sensory deprived "readers." Monks use deep meditation to escape their ego. Psilocybin facilitates the same escape.

Whether used together, individually, or not at all, the ability of meditation and psychedelics to break the tyranny of the ego-oriented default mode network provides vital insight. The mind is more than the material brain. It can even connect with the mind of God.

Psi/extrasensory perception suggests that the mind can unconsciously connect with other minds in a connected universe. Meditation demonstrates that practiced mindfulness can make that connection conscious. Psilocybin blows the door open, allowing a melding of individual and cosmic minds.

We Have Met Additional Blueprints for Human Consciousness

The brain
Neural networks
Electromagnetic fields
The connectome
Mathematics, the mind's eye
The default mode network
Psychedelic receptors

Supplement Twenty-Five

Review the definition and meaning of a "mystical experience." Mystical experiences can be produced by meditation and treatment with psychedelics. They can also occur spontaneously. Whatever their origin, their features are identical.

CHAPTER TWENTY-SIX

The Reality of the Soul

All three sections of this book have arrived at the same conclusion: mind is the master of matter. In the first few moments of the universe, God converted critical equations into fields and forces and executed the creation of electrons and quarks. Mass-encoded quarks and electrons were destined to form stars, a table of elements, and the chemistry of life. Here's a reminder to our materialist colleagues: matter is a derivative product, hardly the primordial substance.

Non-material genetic and protein sequences became material genes and proteins that steered the journey of life. Mutating genes delivered instructions for the future in the DNA of an ancient fish. Later in life's unfolding, untold numbers of genes devoted to the development of neurons supercharged the development of the human brain, creating the most complex material object in the universe.

How do we view the relationship between mind and the brain in this third and final section? Created in God's image, each one of us has received a remarkable gift, a facsimile of the mind of our Creator. Even more remarkable, that mind defines our personhood. Children develop language and personality. Human beings form memories that will last a lifetime. How does the material brain facilitate this near-miraculous process?

The Rest of the Story

When we examine the brain today, we are enamored with neurons and synapses, electromagnetic waves on EEGs, and images of blood flow produced by functional MRIs. Soon, we will see that brain in an entirely different way. Thanks to the Human Connectome Project,[550] we will begin to recognize the brain as a puzzle with trillions of pieces. Each piece is mapped to individual units of mind.

Someday, perhaps, a talented neuroscientist will visit the mathematical landscape and discover God's math for this mapping. We will then recognize the brain as scaffolding for conscious experience and a riverbed for mental flow. The material brain nurtures and hosts our mental experience in a brain-mind partnership that lasts a lifetime.

The near-death experience, terminal lucidity, and psychedelic revelation of the cosmic mind fill in the rest of the story. When the partnership dissolves at the end of life, the mind flows freely on its own. Does that soul retain our personal image? Life reviews are intensely personal,[551] and Max Planck, in at least one NDE, continues to teach quantum mechanics in a "kingdom of light."[552]

Question:

Does matter acting upon matter seem a sufficient explanation for the miracle of mind?

---

[550] Sean P. Fitzgibbon et al., "The Developing Human Connectome Project (dHCP): Automated Resting-State Functional Processing Framework for Newborn Infants," *Neuroimage* 223: article 117303, https://doi.org/10.1016/j.neuroimage.2020.117303.

[551] Bruce Greyson, *After: A Doctor Explores What Near-Death Experiences Reveal about Life and Beyond* (New York: St. Martin Essentials, 2021).

[552] Andrew Macmillan Greyson, in discussion with the author, September 1998.

Substance Dualism

How then do we see the relationship between the mind and the brain? We agree with the idealists that *materialism is baloney.*[553] But we appreciate our embodied time on Earth as a God-given opportunity to learn and mature and personalize our mind. We thus embrace the concept of "substance dualism," a partnership between mind and matter.[554] As expressed by a neuroscientist, "The mind uses the brain, and the brain responds to the mind. The mind also changes the brain. People choose their actions—their brains do not force them to do anything. Yes, there would be no conscious experience without the brain, but experience cannot be reduced to the brain's actions."[555]

A Brief History of Dualistic Thought

This "substance dualism" recognizes that there are two aspects to reality: mind and matter. They interact, but neither can be reduced to the other. This is hardly a unique take on the subject. How does our substance dualism compare to dualist traditions of the past?

From Ancient Greece to the Enlightenment

Plato established a school for higher learning in ancient Greece in 387 BC, and his school remained open to students for nine hundred years. In AD 529, the Christian emperor Justinian I condemned it

---

[553] Bernardo Kastrup, *Why Materialism is Baloney* (Summit, PA: Iff Books, 2014).
[554] Tim Crane and Sarah Patterson, eds., *History of the Mind-Body Problem* (London: Routledge, 2000), 1–2.
[555] Caroline Leaf, "How Are the Mind and the Brain Different? A Neuroscientist Explains," Mindbodygreen, March 8, 2021, https://www.mindbodygreen.com/articles/difference-between-mind-and-brain-neuroscientist.

and closed it for being "pagan."[556] Ironically, Plato distrusted the mental world of the ordinary man. In his form of dualism, the true mind arose from an eternally pure world of "Forms" (ideas) occupying a higher dimension.[557] He considered those forms incorruptible, in contrast with the easily corrupted mind of human beings. Plato's solution for corruptible human minds? Education of an elite class of philosopher-kings.[558] Plato, it is fair to argue, shed little light on the relationship of the mind and the brain.

If Plato distrusted the powers of mind and looked for truth on a higher dimension, René Descartes brought the debate back to Earth. He proposed a metaphysical dualism that radically distinguished between the mind and the body. Deeply religious, he believed that an immortal soul survived the death of the body.[559]

Critics challenged his dualism, noting that thoughts could produce bodily movement. Descartes held his ground, declared his belief, as described by others as that mind and body "bear a relation of act and potency that results in one, whole and complete substantial human being."[560]

Philosophers might prefer the relationship of mind and body as espoused by Thomas Aquinas. Unlike Descartes, Aquinas does not consider the mind a separate substance. A human being is *both* mind and body in unity: "One cannot sense without the body,

---

[556] Paul Kalliga et al., eds. *Plato's Academy: Its Workings and Its History* (Cambridge: Cambridge University Press, 2020).

[557] Stephen Watt, introduction to *Republic*, by Plato, trans. and ed. John Llewelyn Davies and David James Vaughan (London: Wordsworth Editions, 1997), xiv–xvi.

[558] C. D. C. Reeves, *Philosopher Kings: The Argument of Plato's Republic* (Princeton: Princeton University Press, 1988).

[559] Desmond M. Clarke, *Descartes, A Biography* (Cambridge: Cambridge University Press, 2006).

[560] Justin Skirry, "René Descartes: The Mind-Body Distinction," Internet Encyclopedia of Philosophy, https://iep.utm.edu/descartes-mind-body-distinction-dualism/.

therefore the body must be some part of man."[561] For Aquinas, humans are *both* body and soul. The distinction may seem arbitrary, since Aquinas clearly believes the soul will survive eternally.

Our conclusion is Cartesian. Mind maps to the brain, and the brain personalizes the mind. The mind survives death of the body. *God gave human beings a soul.*

Judeo-Christian Tradition

The Judeo-Christian Bible declared long ago that human beings are both physical and spiritual beings. In Deuteronomy 11:18, NIV, Moses commands that the people "[f]ix these words of mine in your hearts and minds; tie them as symbols on your hands and bind them on your foreheads."

In Psalm 26:2 (NIV), the psalmist declares, "Test me, Lord, and try me, examine my heart and my mind." Note that no psalmist suggests, "Examine my brain."

In Mark 8:36–37 (NIV), Jesus asks the rhetorical question, "What good is it for someone to gain the whole world, yet forfeit their soul?"

In Matthew 22:36–37 (NIV), He declares the greatest commandment: "Love the Lord your God with all your soul and with all your heart and with all your mind."

In John 4:24, (NIV) Jesus tells us that "God is spirit, and his worshipers must worship in the Spirit and in the truth." Worship, Jesus tells us clearly, is primarily an act of mind. In the Judeo-Christian tradition, the soul is the inner person destined for eternal life.

---

[561] Paul Chutikorn, "A Thomistic Critique of Cartesian Dualism," Thomistica, November 9, 2018, https://thomistica.net/essays/2018/11/9/a-thomistic-critique-of-cartesian-dualism.

More than 150 biblical references focus on the mind. A sampling of these verses demonstrates that God cares very much about what happens in our minds and thoughts. The Apostle Paul emphasizes this concern. "Those who live according to the flesh have their **minds** set on what the flesh desires; but those who live in accordance with the Spirit have their **minds** set on what the Spirit desires. The **mind** governed by the flesh is death, but the **mind** governed by the Spirit is life and peace." Romans 8:5–6 (NIV).

Nondualism

Not all faiths are based on dual substance. Some of the world's oldest religions take the opposite view, considering dualism an illusion. The distinction between matter and mind or brain and consciousness is a temporary Earthbound convenience. The purpose of life from this perspective is to see "dual substance" as a false dichotomy.[562]

Buddhism

In Buddhist belief, the body is considered a guest house and the mind an indwelling guest. When individuals die, according to the Buddhist concept of reincarnation, the mind leaves the body to be reborn in another. The guest leaves one home and moves to another.

Buddhists can spend an entire life investigating the nature of consciousness through meditation. Obeying an ancient text, translated to the Discourse on the Establishing of Mindfulness, Buddhists cultivate mindfulness to "overcom[e] sorrow and lamentation, for

---

[562] Matthieu Ricard and Wolf Singer, *Beyond the Self: Conversations between Buddhism and Neuroscience* (Cambridge. MA: MIT Press, 2017).

the destruction of suffering and grief, for reaching the right path, for the attainment of Nirvana."[563]

Hinduism

Focusing on an imbalanced "ego" burdened with attachments and opinions, Hinduism focuses on a true "self" stripped of attachments and desires. Through meditation, Hindus hope to experience God as an indwelling spirit. Hinduism, then, embraces a unique form of dualistic belief. Human beings have mind, considered a form of energy by the Hindu faith. That mind is accompanied by nothing less than the indwelling spirit of God.[564]

In the documentary *The Elephant Whisperers*, viewers are graced by the story of an indigenous people nurturing injured baby elephants back to health.[565] The emotional intelligence of the elephant is on wondrous display, as is the happiness and kindness of their keepers. One brief scene includes views of an outdoor Hindu shrine dedicated to offerings and prayer. One can only believe that God looks on these gentle people and their mind-God dualism and sees it as very good.

---

[563] Dalai Lama and Thubten Chodron, Samsara, Nirvana, and Buddha Nature (Somerville, MA: Wisdom Publications, 2019).

[564] Mohan R. Pandey, *Hinduism: A Path to Inner Peace* (self-pub., CreateSpace, 2013).

[565] Pallavi Keswani, "'The Elephant Whisperers' Documentary Review: A Strikingly-Lush Safari on the Co-Existence of Man and Nature," *The Hindu*, last updated December 28, 2022, https://www.thehindu.com/entertainment/movies/the-elephant-whisperers-documentary-review-strikingly-lush-safari-on-coexistence-of-man-and-nature/article66310108.ece.

Mind and Brain in Materialism

As confusing as this brief review of dualistic thought might seem, each of these well-known positions seems superior to modern materialist theory. The mainstream material position that the mind is nothing more than the brain is nonsensical. Materialists hold that matter is the fundamental substance in nature, and all mental states, all aspects of consciousness, are produced by the interaction of matter with—wait for it—additional matter. Somehow neurons, neurochemicals, and electromagnetic fields produce the green of grass, the taste of caramel, and the feelings of remorse and joy.

That same matter, in this worldview, produces a terminal illusion called the "near-death experience" to soften the blow of death. The same mind that discovers the math of space-time and black holes, the same soul that hungers for love and meaning and a relationship with God, is no more permanent than the grin of the Cheshire Cat.

After several thousand years of thought and science, materialists (prophets of "scientism" who see matter as the primordial substance) have expunged the concept of an independent mind. They might quote *Macbeth*: "Life's but a walking shadow, a poor player that struts and frets his hour upon the stage, and then is heard no more. It is a tale told by an idiot, full of sound and fury, signifying nothing."[566]

Thomas Aquinas, René Descartes, Jesus Christ, and the devotees of Eastern religions would find that a strange conclusion.

---

[566] William Shakespeare and Jonathan Drain, *The Tragedy of McBeth*, 2017, Kindle.

Supplement Twenty-Six

How do we define a miracle? Are they still possible in the modern world?

CHAPTER TWENTY-SEVEN

The Reality of God

Science and faith are complimentary revelations of God.

The Consonance of Science and Faith

The journey of science is far from over. Cosmologists speculate about dark energy,[567] dark matter,[568] and qubits as the nodes of space-time.[569] Scientists pursue quantum computers[570] and explore protein folding with artificial intelligence.[571] Biologists continue to explore the genome, while physicians plan CRISPR-based gene treatments.[572]

Science is a story of evolving theories as paradigms slowly shift. That makes the faith-based explanation for the relationship of the mind and the brain even more remarkable for the consistency of

---

[567] Brian Clegg, *Dark Matter and Dark Energy: The Hidden 95% of the Universe* (London: Icon Books, 2019).

[568] Priyamvada Natarajan, *Mapping the Heavens: The Radical Scientific Ideas That Reveal the Cosmos* (New Haven: Yale University Press, 2017), 96.

[569] Seth Lloyd, *Programming the Universe: A Quantum Computer Scientist Takes on the Cosmos* (New York: Vintage Books, 2007), 70.

[570] Chris Bernhardt, *Quantum Computing for Everyone* (Cambridge, MA: MIT Press, 2020).

[571] Robert F. Service, "The Game Has Changed. AI Triumphs at Protein Folding," *Science* 370, no. 6521 (December 2020): 1144–1145, https://doi.org/10.1126/science.370.6521.1144.

[572] Kevin Davies, *Editing Humanity: The CRISPR Revolution and the New Era of Genome Editing* (New York: Pegasus Books, 2020).

its message. Mind—not nature—created the universe. The Mind (capital *M*) of our Maker explains the human mind (lower case *m*), complete with our quest for quarks and equations. A conflict between science and faith is an illusion.

More than any prior generation of humanity, we can understand and celebrate God's creative power. We are the recipients of transcendent breakthroughs in cosmology, gene sequencing, and neuroscience only available in the past few decades. In some cases, scientific breakthroughs are but a few years old. Our explanation for human consciousness remains steadfast: we are created in the image of God.

Two Books

God has revealed Himself in two ways: Holy Scripture, written by many authors over centuries, tells the story of His connection with human minds. The book of His works, under constant edit and expansion, celebrates the complexity and ingenuity of His creation.

Do these books paint contradictory portraits, or do they provide mutually reinforcing narratives? We should test some fundamental attributes of God and see if they resonate with both books.

The ancient Greek philosopher Anaxagoras of Clazomenae proposed that the sun was a fiery rock instead of a solar deity.[573] For this impropriety, the Athenians sentenced him to death. He remains well known for a second proposal: mind (*nous* in Greek) initiated and governs the cosmos. Two-and-a-half millennia later, science is confirming his proposals. Yes, the sun is a fiery rock

---

[573] Daniel W. Graham, *The Texts of Early Greek Philosophy: The Complete Fragments and Selected Testimonies of the Major Presocratics, Part 1* (Cambridge: Cambridge University Press, 2010), 314.

(actually, a collection of fiery gases), and yes, an unimaginable mind created and supervises the cosmos.

> **Question:**
> *How do human beings remember that love, a component of sacred science, reflects the ultimate priority of God?*

Compatibility Between Science and Scripture

The following aspects of reality are shared by both science and faith.

1. The universe erupted from nothingness.
2. Much of the created order is immaterial.
3. Mathematics is the foundation of creation.
4. Should God exist, He is outside the created order, unaffected by distance or time, but intimately entangled with His creation.
5. Life unfolded as a succession of species.
6. Human beings are distinct from all other species: we have the ability to reason and the desire to create.
7. The sun took shape in the last third of creation's history.
8. Faith is a vital component of culture.
9. Love is the fuel of creation.

God Created the Universe in a Flash of Light

As recently as the 1950s, leading scientists thought our universe was eternal. Einstein rejected the concept of a "big bang," and astronomer Fred Hoyle called the idea "anti-scientific."[574] Thanks to the measurement of light elements in the universe, the detection of the cosmic background radiation, and, even more, the recognition that our universe is expanding, we know vastly more today about our beginnings than the leading men of science at the middle of the twentieth century. "And God said, let there be light!"[575]

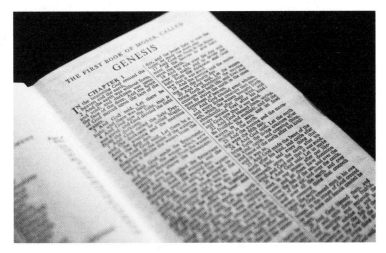

While the Bible is not a scientific textbook, ancient authors were inspired to describe creation out of nothingness (creatio ex nihilo) in an explosion of light. Of the six days of creation, what day did the sun appear? Day four!! What light were they describing at time zero?

Images from Hubble and the James Webb Space Telescope put the magnificence of our universe on full display, confirming the reality of the big bang.

---

[574] Paul Halpern, *Flashes of Creation: George Gamow, Fred Hoyle, and the Great Big Bang Debate* (New York: Basic Books, 2021).
[575] Genesis 1 (NIV).

Much of the Created Order is Immaterial. Do Science and Faith Agree?

Jesus tells us that God is spirit:[576] God has no body and is omnipresent.

1. Without the Higgs field, the entire universe would be equivalent to spirit, a vast space devoid of physical structure.

2. Quantum particles are more spirit than matter, defying conventional concepts of location and time.[577] Atoms are 99 percent empty space.[578]

3. How much of our universe can classify as physical? Just 4 percent.[579] The rest is immaterial, including hypothetical dark matter and dark energy. Not only is God spirit but also is 96 percent of His creation.

Mathematics Provides the Foundation for the Created Order. Can Mathematicians Refute This Assertion?

We have noted repeatedly that math carries intimations of the divine. We should repeat the words of journalist Jerry Bowyer in his summary of New Testament professor Vern Poythress's argument about mathematical formulas: "They are true everywhere (omnipresent), true always (eternal), cannot be defied or defeated

576 John 4:24 (NIV).
577 Jim Al-Khalili, *Quantum: A Guide for the Perplexed* (London: Weidenfeld and Nicolson, 2012), 24.
578 Richard Reeves, *A Force of Nature: The Frontier Genius of Ernest Rutherford* (New York: W. W. Norton and Company, 2007), 40.
579 Clegg, *Dark Matter and Dark Energy*.

(omnipotent), and are rational and have language characteristics (which makes them personal). Omnipresent, omnipotent, eternal, personal… Sounds like God."[580]

In this worldview, God began creation with blueprints, a library of mathematical equations. Those equations are eternal and nonphysical—God's thoughts before the beginning of time. A mathematical library in a higher dimension is the obvious explanation for the "unreasonable effectiveness of mathematics in the natural sciences."[581]

God is Unlimited by Distance or Time. Does Science Suggest This is Possible?

In 1905, Max Planck and Albert Einstein launched the field of quantum mechanics.[582] Subsequent studies of the quantum domain have produced one inexplicable finding after another.

During their travels, fundamental particles proceed as probability equations, exploring multiple paths in the same instant, only one of which will lead to its actual destination.[583] Waves can entangle with each other and function as a single system, even across great distances of space.

Today, we live in a world awash with quantum information theory, waiting for Google, Microsoft, or one of their competitors

---

[580] Jerry Bowyer, "God in Mathematics," *Forbes*, April 19, 2016, https://www.forbes.com/sites/jerrybowyer/2016/04/19/where-does-math-come-from-a-mathematiciantheologian-talks-about-the-limits-of-numbers/.

[581] Eugene P. Wigner, "The Unreasonable Effectiveness of Mathematics in the Natural Sciences," *Communications on Pure and Applied Mathematics* 13, no.1 (1960): 1–14, https://doi.org/10.1002/cpa.3160130102.

[582] J. L. Helibron, *The Dilemmas of an Upright Man: Max Planck and the Fortunes of German Science* (Cambridge: Harvard University Press, 2000), 404.

[583] Al-Khalili, *Quantum*.

to produce a functional quantum computer.[584] Feynman's statement remains true today: "Nobody understands quantum mechanics."[585] But people of faith can recognize quantum superposition and entanglement as divinely commissioned powers, impervious to the constraints of classic distance and time.

Einstein could make a similar argument. In the theory of relativity, both time and distance are dependent on the observer's frame of reference. Who are we to limit God to a single frame?

God is Entangled with All of Creation. Does Science Affirm This Possibility?

In Matthew 6:6–8 (NIV), Jesus informs us: "And when you pray, do not keep on babbling like pagans, for they think they will be heard because of their many words. Do not be like them, for your Father knows what you need before you ask him."

Psalm 139 (NIV) delivers the same message: "Where can I go from your Spirit? Where can I flee from your presence? If I go up to the heavens, you are there; if I make my bed in the depths, you are there."

Leading theoreticians are proposing that quantum logic gates (qubits) may tie the nodes of space-time together.[586] Others suggest

---

584 John Russell, "IBM Quantum Update: Q System One Launch, New Collaborators, and QC Center Plans," HPCwire, January 10, 2019, https://www.hpcwire.com/2019/01/10/ibm-quantum-update-q-system-one-launch-new-collaborators-and-qc-center-plans/.

585 Richard Feynman, "Probability and Uncertainty: The Quantum Mechanical View of Nature," 1964, in Messenger Lectures at Cornell University, California Institute of Technology and The Feynman Lectures Website, transcript and HTML5 video, at 8:10, 56:32, https://www.feynmanlectures.caltech.edu/fml.html#6.

586 Lloyd, *Programming the Universe*.

that the universe is holographic.[587] Do these quotes from the Gospel of Matthew and the Psalms support the concept that the universe is a quantum information device broadly entangled with its Creator?

The Apostle Paul,[588] modern near-death survivors,[589] and Christ himself remind us to expect a life review.[590] It is hard to escape the conclusion: every aspect of the created order has a direct connection with God.

In Genesis 1, God Commissioned a Succession of Living Creatures. Is a Library for Life Compatible with a Divine Role in Life's Journey?

And God said, "Let the water teem with living creatures.... Let the land produce living creatures according to their kinds: the livestock, the creatures that move along the ground, and the wild animals, each according to its kind." And it was so.[591]

Atheistic skeptics have embraced a distorted and soon-to-be-outdated model of biological evolution that denies God any role in the process. They have described life's journey as an unguided product of random mutation possessing no direction or purpose.[592] The study of plant and animal genomes in just the past few years is telling a different story. Neutral mutations may visit an eternal

---

[587] Anil Ananthaswamy, "Is Our Universe a Hologram? Physicists Debate Famous Idea on its 25th Anniversary," *Scientific American*, November 30, 2022, https://www.scientificamerican.com/article/is-our-universe-a-hologram-physicists-debate-famous-idea-on-its-25th-anniversary/.

[588] Romans, 2:6 (NIV).

[589] Bruce Greyson, *After: A Doctor Explores What Near-Death Experiences Reveal about Life and Beyond* (New York: St. Martin Essentials, 2021).

[590] Matthew 12:36 (NIV), 2:6 (NIV).

[591] Genesis 1:20–25 (NIV).

[592] Stephen Jay Gould, *The Structure of Evolutionary Theory* (Cambridge, MA: Belknap Press, 2002).

library and retrieve gene sequences for the future.[593] The non-coding portion of the genome provides safe storage for those genes, in a process that explains multiple mysteries of the journey of life:

1. The abrupt appearance of new species in the ancient oceans[594]
2. Genes for life on land in the genome of ancient fish[595]
3. The multispecies acquisition of echolocation[596]
4. The blossoming of human consciousness worldwide[597]

Only One Creature Shares His Image

"Then God said, 'Let us make mankind in our image, in our likeness, so that they may rule over the fish in the sea and the birds in the sky, over the livestock and all the wild animals, and over all the creatures that move along the ground.' So God created mankind in his own image, in the image of God he created them; male and female he created them."[598]

Science would agree that human beings alone have higher consciousness, complete with a spiritual instinct, a hunger for meaning, an urge to create, and the ability to contemplate eternity. John Lennox, Northern Irish mathematician, Christian apologist, and

---

[593] Andreas Wagner, *Arrival of the Fittest: Solving Evolution's Greatest Puzzle* (New York: Current, 2014).

[594] Simon Conway Morris, *The Crucible of Creation: The Burgess Shale and the Rise of Animals* (Oxford: Oxford University Press, 1999).

[595] Elizabeth Pennisi, "Genes for Life on Land Evolved Earlier in Fish," *Science* 371, no. 6530 (February 2021): 658–659, https://doi.org/10.1126/science.371.6530.658.

[596] Fredric M. Menger, "An Alternative Molecular View of Evolution: How DNA was Altered over Geological Time," *Molecules* 25, no. 21 (November 2020): article 5081, https://doi.org/10.3390/molecules25215081.

[597] Fredric M. Menger, "Molecular Lamarckism: On the Evolution of Human Intelligence," *World Futures* 73, no. 2 (May 2017): 89–103, https://doi.org/10.1080/02604027.2017.1319669.

[598] Genesis 1:26–28 (NIV).

professor (emeritus) of mathematics at Oxford University, gives us basic reminders: "Not only did we not create the universe, but we did not create our own powers of reason either. We can develop our rational faculties by use; but we did not originate them. How can it be, then, that what goes on in our tiny heads can give us anything near a true account of reality? How can it be that a mathematical equation thought up in the mind of a mathematician can correspond to the workings of the universe?" He continues: "The Bible gives us a reason for trusting reason. Atheism does not. This the exact opposite of what many people think."[599]

Reflecting the image of the ultimate creative power, the divine Logos, human beings have an innate creative drive. The history of science and technology tells a never-ending story of humankind's quest for innovation.

The Sun Appeared After Two-Thirds (on the Fourth of Six Days) of Creation. Does Science Support That Timeline?

Inspired scripture provides powerful evidence for the reality of God. Have you attempted to outline six days of creation? Did you begin with an explosion of light? Did you wait until the fourth of six days to create the sun, the source of Earth's light?

Ancient Egyptians worshipped the sun as a god. Living shoulder to shoulder with Egyptians, ancient Israel reached a different conclusion. The first chapter of Genesis tells us that the one true God hung the sun on day four of six, two-thirds of the way through creation.[600]

---

[599] John C. Lennox, *Can Science Explain Everything?* (Charlotte: The Good Book Company, 2019).
[600] Genesis 1:16 (NIV).

Science tells us that the universe is 13.8 billion years old. The sun is 4.6 billion years old. The difference in their ages is 9.2 billion years. If we take the proportion of time between the beginning of time and the creation of the sun and divide that number by the accepted age of the universe, the result (9.2/13.8) is .67—the equivalent to the fourth of six days.

Genesis 1 is not a scientific textbook. But when it comes to the origin of the universe and the creation of the sun, science and Genesis 1 are in remarkable agreement. As previously noted by the physicist and Anglican priest John Polkinghorne: "If people in this so-called 'scientific age' knew a bit more about science than many of them actually do, they'd find it easier to share my view." That view: "science and faith are complementary paths to a fuller understanding of God."[601]

The Cultural Impact of Faith

Monotheism took root in ancient Israel as God freed his chosen people from slavery, provided manna in the wilderness, issued the Ten Commandments, and ordered that "[when] you reap the harvest of your land, do not reap to the very edges of your field or gather the gleanings of your harvest. Leave them for the poor and for the foreigner residing among you."[602]

God inspired messianic dreams and the production of Holy Scripture. Then the life and message of Jesus Christ changed our method of dating from "before Christ" (BC) to "Anno Domini," or "in the year of our Lord," (AD). Christianity emerged in the

---

[601] John Polkinghorne, *Quarks, Chaos and Christianity: Questions to Science and Religion* (Chestnut Ridge, Crossroad, 1994), xii.
[602] Leviticus 23:22 (NIV).

Roman Empire at a perfect time in world history and the history of temple worship.

Jesus understood human nature as only the Creator might. Yes, we would store up riches, look with lust in our heart, avoid the fallen Samaritan, make a show of our prayer, hate our enemies, and disappoint God over and over. But Jesus brought reconciliation.

The message of Jesus was pitch-perfect, his parables profound, his healing miraculous, his fulfillment of prophecy undeniable, and his message clear and concise: love God with all your heart, strength, soul, and mind, and love your neighbor as yourself.

The Judeo-Christian story has had an inestimable impact on the subsequent trajectory of culture. The story of resurrection converted an empire, bridged ancient Greece with the Renaissance, and helped launch the journey of science. Newton was the first to describe a force of nature in mathematical terms.[603] Did he believe his ability to describe gravity in an equation reduced his dependence on God? To the contrary, he attributed that equation to the God of his faith. Imagine how Newton would react to the full repertoire of God's creative blueprints identified by science today. What would he think of $E = mc^2$, the code of up and down quarks, the Higgs mechanism, quantum mechanics, and strong and weak nuclear forces?

---

[603] James Gleick, *Isaac Newton* (New York: Vintage Books, 2003), 110.

Love is the Fuel of Creation and the Number One Priority for God. Is There Scientific Evidence to Support This Assertion?

Orphans deprived of a mother's love have underdeveloped brains and lifelong difficulties with relationships.[604]

As mentioned, atheists treated with psilocybin encounter a loving force they describe as "conscious, benevolent, intelligent, sacred, eternal, and all-knowing."[605] As a result, patients escape depression and nicotine addiction, and a majority who considered themselves atheists develop a new belief in God.[606]

In response to psilocybin, neurons develop more dendrites and create new synapses. Whether the source is a mother's love or cosmic love, love is neuroplastic, described earlier as a growth factor for the human brain.[607]

Conflict Between Science and Faith: Total Myth

There is an enormous irony at work in modern culture. Scientism (as distinct from science) has declared itself the voice of reason, an

---

[604] Shaun Walker, "Thirty Years On, Will the Guilty Pay for Horror of Ceaușescu Orphanages?" *The Guardian*, December 15, 2019, https://www.theguardian.com/world/2019/dec/15/romania-orphanage-child-abusers-may-face-justice-30-years-on.

[605] Roland R. Griffiths et al., "Survey of Subjective 'God Encounter Experiences': Comparisons Among Naturally Occurring Experiences and Those Occasioned by the Classic Psychedelics Psilocybin, LSD, Ayahuasca, or DMT," *PLoS One* 14, no. 4 (2019): 1–26, https://doi.org/10.1371/journal.pone.0214377.

[606] Guy M. Goodwin et al., "Single-Dose Psilocybin for a Treatment-Resistant Episode of Major Depression," *New England Journal of Medicine* 387, no. 18 (2022):1637–1648, https://doi.org/10.1056/nejmoa2206443.

[607] Maxemiliano V. Vargas et al., "Psychedelics Promote Neuroplasticity through the Activation of Intracellular 5-HT2A Receptors," *Science* 379, no. 6633 (February 2023): 700–706, https://doi.org/10.1126/science.adf0435.

unstoppable, rational force.[608] Science, it claims, will single-handedly reveal the secrets of the universe and defeat humankind's hunger and disease. Science will bring heaven to Earth and meet all our hopes and needs.

But one human need will go unfulfilled: the need for meaning. One hope will go unmet: the hope for an eternal future. Only God could fulfill those hopes and needs. Why would we ever let a prideful scientism jeopardize a faith-based understanding of the world?

Valued Science

We have welcomed a remarkable team of talented and well-intentioned individuals to the diverse fields of science. Their work enriches our lives and expands our understanding of the universe. Science is a noble and meaningful human endeavor. What better way can we honor our God than by studying and celebrating His scientific works?

But we fatigue of the tiresome and fully discredited trope that "science conflicts with faith." Science reaffirms the message of scripture, and John Lennox summarizes it well: "Far from science having buried God, not only do the results of science point towards his existence, but the scientific enterprise itself is validated by his existence."[609]

Human beings always have a choice. We can view the world as an accidental product of a transient, material nature, short on meaning but our only source of material possessions and physical comfort. Or, we can see the world as a gift from God where meaning

---

[608] Thomas Burnett, "What is Scientism?" American Association for the
       Advancement of Science, May 21, 2012, https://sciencereligiondialogue.org/
       resources/what-is-scientism/.
[609] John C. Lennox, *God's Undertaker: Has Science Buried God?* (Oxford, England:
       Lion Books, 2009).

transcends matter. The first is a world where the government takes precedence over the individual and discourages the practice of religion (sad to say, like modern-day China). The latter is a world devoted to the sanctity of the individual that protects freedom of religion as a vital component of mental and physical well-being.

Faith inspires us to acknowledge our Creator. Science describes the process of His creation. What science conflicts with faith? Only science incorrectly interpreted. What faith denies our science? Only faith prematurely defeated.

Reality does not come to us divided into two parts: the first from ancient and irrelevant scrolls, the second from colliders and computers. Scripture provides a foundational view of the Master of the Heavens and the Earth. Science enriches our view of that Master by demonstrating the scale of His mastery.

Celebrating the Book of God's Works

This book ends with a gentle admonition to scientists: your work is honorable and provides an invaluable service to humankind. As you toil in your labs, modify your equations, or design the next space telescope, keep this thought in mind: people of faith believe you are walking on hallowed ground.

It also ends with an admonition to people of faith: just as Joshua crossed the Jordan and took possession of the Promised Land, we should embrace the Book of God's Works.

Understanding its contents will strengthen and deepen your faith. Yes, that book is full of science, but it is science that promotes the most valid worldview. Embracing sacred science is an act of loving God with all our mind.

If you believe in God, you believe He created every quark and electron, every field and force, and every species. If you believe in the Maker of the Heavens and the Earth—and many people of faith affirm that belief on a regular basis by reciting the Apostles' Creed[610]—you acknowledge Him as author of the universe and life. If that faith is genuine, you should reclaim science for your faith. The source of that science is sacred!

Let Us Relist the Blueprints of Consciousness and Add The Most Powerful Blueprint of Them All:

The brain
Neural networks
Electromagnetic fields
The connectome
Mathematics, the mind's eye
The default mode network
Psychedelic receptors
Love

Yes, love. For we are one. This is the teaching of sacred science.

---

[610] *Dictionary of the Christian Church*, s.v. "Apostles' Creed" (Oxford: Oxford University Press, 2005), 90.

Supplements

The First Twenty Minutes

In 1905 Belgium, ten-year-old George Lemaître made two momentous decisions:[611] First, he would become a priest. Second, he would study physics. Combining the two, he would launch a rare career—he would become an astronomer-priest. He postponed his dreams between 1914 and 1918 while serving as an artillery officer in WWI, receiving the Belgian War Cross. During the war, he used every spare moment to devour books on mathematics and physics.

Following the war, he received a degree in physics in 1920 and Catholic ordination in 1923, all the while digesting the mathematics of Einstein's radical new view of the universe, the general theory of relativity, published in 1915

Understanding Einstein's Equations

Lemaître saw something in Einstein's equations that no one else had noticed: Einstein had added a fudge factor to the equations of relativity called the "cosmological constant." His constant served a vital

---

[611] Priyamvada Natarajan, *Mapping the Heavens: The Radical Scientific Ideas That Reveal the Cosmos* (New Haven: Yale University Press, 2017).

purpose: it produced an eternally stable universe.[612] Most scientists at the time believed the size of the universe would never change.

To describe an unchanging universe, Einstein needed a constant in his equations that would counteract the gravitational pull of matter that might cause the universe to collapse.[613]

Lemaître had a moment of insight: remove that cosmological constant, and the universe might be expanding. If the universe is expanding and you rewound the video of its journey, all that energy and matter would collapse to a moment of creation. Lemaître proposed that the world was born in a tiny point—a "primeval atom" or a "cosmic egg."[614] This meant space and time had a beginning.

Abominable Physics

In a famous encounter in 1927, Einstein reacted to the big bang concept: "Your calculations are correct, but your physics is atrocious."[615]

---

[612] Walter Isaacson, *Einstein: His Life and Universe* (New York: Simon and Schuster, 2007), 254.

[613] Isaacson, *Einstein*, 353.

[614] Paul Halpern, *Flashes of Creation: George Gamow, Fred Hoyle, and the Great Big Bang Debate* (New York: Basic Books, 2021), 63.

[615] Ross Pomeroy, "Why Georges Lemaître Should be as Famous as Einstein," Real Clear Science, March 7, 2017, https://www.realclearscience.com/blog/2017/03/07/why_georges_lemaitre_should_be_as_famous_as_einstein.html.

The Catholic priest Lemaitre first recognized that the cosmological constant inserted into the equations of Relativity was a mistake. Remove the constant, Lemaitre realized, and the universe could be expanding. Einstein's reaction: "Your math is correct, but your physics is abominable!" Hubble and his telescope soon confirmed Lemaitre's observation and his hypothesis that the universe began as a "cosmic egg."

But in 1929, Edwin Hubble and his deep-space telescope reported that galaxies were moving away from each other like raisins in a baking loaf of bread. Run the movie in reverse, Hubble realized, and those stars once existed in a single atom.[616] By 1933, Einstein had changed his tune. "This is the most beautiful and satisfactory explanation of the creation to which I have ever listened."[617]

[616] Gale E. Christianson, *Edwin Hubble: Mariner of the Nebulae* (Boca Raton, Florida: CRC Press, 1997).
[617] Simon Singh, *Big Bang: The Origin of the Universe* (New York: Harper Perennial, 2005), 276.

He also considered the cosmological constant the greatest blunder of his career.[618]

The physicist Fred Hoyle resolutely dismissed the idea of a sudden origin of the Universe as unacceptable on both scientific and philosophical grounds.[619] Considering a beginning for the universe a joke, Hoyle gave it a derogatory name, the "big bang."[620] But the name big bang resonated with the public, and soon no one was laughing.

Religious Claim

Pope Pius XII was quick to claim the big bang theory as scientific evidence for a God-created universe.[621]

Fearing that his scientific colleagues would dismiss his idea because of its religious implications, Lemaître resisted the Pope's claims, insisting that there was neither a connection nor a conflict between his religion and his science. He believed that religion and science were separate and valid views of the world, providing different, parallel interpretations of reality.[622]

[618] Singh, *Big Bang*, 274.
[619] Helge Kragh, *How did the Big Bang get its name? Here's the real story.* Nature 627, 726-728 (2024).
[620] Halpern, *Flashes of Creation*, 12.
[621] Pomeroy, "Why Georges Lemaître."
[622] Singh, *Big Bang*, 362.

The Higgs Field

No blueprint is more significant than the Higgs field (and Higgs bosons and the Higgs mechanism that delivers mass).

The most common attempt to describe the Higgs mechanism involves the following metaphor: a famous individual enters a crowded room.[623] As inhabitants of the room move toward their celebrated guest, he or she has difficulty navigating the gathering crowd. The famous visitor slows down as though acquiring incremental mass.

We might use a metaphor of water skiing to provide another illustration. Different particles, you might imagine, vary in their water-skiing talent. The photon is a slalom skier, so skillful, it skims above the Higgs Ocean, ignoring its waves. No mass for the agile photon.

Electrons are less talented. Their ski tips dip slightly into the Higgs Ocean and assume a small amount of mass.

We should classify quarks as clumsy skiers, sinking into the ocean and taking on full doses of mass. The ratio of the proton's mass to the electron's must be precisely 1 to 1,836 to produce a

---

[623] Kathryn Jepson, "Famous Higgs Analogy, Illustrated," Symmetry Magazine, September 6, 2013, https://www.symmetrymagazine.org/article/september-2013/famous-higgs-analogy-illustrated?language_content_entity=und.

stable atom—one of the hundreds of fine-tuned parameters in the universe necessary to support the development of life.[624] The Higgs mechanism delivers precisely that quantity of mass.

Now imagine wearing boxing gloves and taking practice swings at imaginary particles. Swing at the photon and miss. Deliver a glancing blow to the electron. Then strike a passing quark full bore. What are you accomplishing in this process? You are encoding mass values for the universe. The Higgs field is the motherboard of creation.

Nobel prize–winning theoretical physicist Frank Wilczek has provided a scientific explanation for the Higgs mechanism.[625] When certain metals approach near-zero temperature, they become "superconductors." In a superconductor, electrons flow in two-electron teams, eliminating electrical resistance. Superconducting technology produces the powerful magnets used in MRI machines and "magnetic levitation" trains that float above the tracks.[626]

Now to the major point: evidence suggests that photons traveling through a superconductor encounter electrons flowing in doublets, slow down, and acquire a small quantity of mass.[627] Wilczek's conclusion: empty space is not empty at all but filled with an invisible superconducting ocean. The Higgs field makes "'empty space'…a super-duper-superconductor."[628]

---

[624] Geraint F. Lewis and Luke A. Barens, *A Fortunate Universe: Life in a Finely Tuned Cosmos* (Cambridge: Cambridge University Press, 2016).

[625] Frank Wilczek, "A Particle That May Fill 'Empty' Space," *Wall Street Journal*, December 29, 2022, https://www.wsj.com/articles/a-particle-that-may-fill-empty-space-11672337133.

[626] Stewart C. Bushong and Geoffrey Clarke, *Magnetic Resonance Imaging: Physical and Biological Principles*, 4th ed. (St. Louis: Elsevier, 2015).

[627] Piers Coleman, "Phillip W. Anderson, (1923–2020)," *Nature*, May 1, 2020, https://www.nature.com/articles/d41586-020-01318-4.

[628] Wilczek, "A Particle."

Implication of the Higgs Equation for "Random Evolution"

In the 1960s, Peter Higgs and other physicists recognized the potential power of the Higgs field by examining its equation.[629] We should give full thought to the significance of that development. God is the ultimate mathematician, and the fundamental equations of physics are His thoughts. Because He created us in His image, we can analyze His equations and understand His math. This human ability poses a lethal challenge to random evolution and the material worldview. Do atheists expect us to believe that evolution randomly gave us the ability to decipher higher math?

We might believe that evolution has given us the ability to dodge rocks, shoot arrows, and scan the ground for snakes. It seems implausible that random evolution produced our ability to explore advanced math. The mind of God wrote the Higgs equation. The human mind, created in His image, comprehended His math.

Deeper Dive: How Did Scientists React to the Higgs Mechanism?

Creators of the standard model hoped to develop a no-God explanation for the fields and forces of the universe. Unified at time zero, the forces of physics unfolded like Russian Matryoshka dolls into separate forces through a process of broken symmetry.[630] They discovered deep beauty in their particle model with its multiple particles,

---

[629] Peter W. Higgs, "Broken Symmetries and the Masses of Gauge Bosons," *Physics Review Letters* 13, no. 16 (October 1964): 508, https://doi.org/10.1103/PhysRevLett.13.508.

[630] Lawrence M. Krauss, *The Greatest Story Ever Told—So Far* (New York: Atria Books, 2017), 217.

fields, and forces. It felt like a stunning defeat to require a strange new field like no other field in science to deliver the missing mass.

Juan Maldacena, the Carl P. Feinberg Professor at the Institute for Advanced Study in Princeton, New Jersey, admits to this ambivalence in an essay entitled "The symmetry and simplicity of the laws of physics and the Higgs boson."[631] "Our present understanding of particle physics," he admits, "is like the story of the Beauty and the Beast. Beauty represents the forces of nature: electromagnetism, the weak force, the strong force, and gravity. These are all based on the symmetry principle, called gauge symmetry. In addition, we also need the beast, which is the so-called Higgs field. It contains much of the mysterious and strange (some would say 'ugly') aspects of particle physics."

Why Consider Higgs "Ugly"?

Given that the Higgs field and its bosons are a necessary code of creation, why describe it as ugly? In ancient Greek plays, playwrights lowered a make-believe God in a mechanical balloon ("deus ex machina") to provide the solution to the play's problems. In the case of the Higgs field, this "god from a machine" delivered mass.

Large Hadron Collider Produces the Higgs Boson

With the discovery of the Higgs mechanism, particle physicists acknowledged that their "beauty" depended on a "beast." Particle physics found its version of manna on the ground. There was no mass, and then a miracle happened!

---

[631] Juan Maldacena, "The Symmetry and Simplicity of the Laws of Physics and the Higgs Boson," European Journal of Physics 37, no. 1 (October 2014): Article 015802, https://doi.org/10.1088/0143-0807/37/1/015802.

The Toilet

Listen to the words of Sheldon Glashow, one of the inventors of the standard model (italics mine):[632]

> "*The Higgs is like a toilet. It hides all the messy details we would rather not speak of. The Higgs field, as elegant as it might be, is within the Standard Model essentially an ad hoc addition.* It is added to the theory to do what is required to accurately model the world of our experience. But it is not required by the theory. The universe could have happily existed with a long-range weak force and massless particles. We would just not be here to ask.... Moreover, the detailed physics of the Higgs is, as we have seen, undetermined within the Standard Model alone. That Higgs could have been twenty times heavier, or one hundred times lighter."

Triumph or Tragedy

Given their discomfort with the Higgs mechanism, how did scientists react to the moment of triumph when the Large Hadron Collider sparked the Higgs field with enough energy to prove the existence of the Higgs boson? They were deeply ambivalent.[633]

Surely, the collider would point the way to a new model of the universe, placing the Higgs in a larger context. Surely, there would

---

[632] Krauss, *The Greatest Story Ever Told*, 276.
[633] Adrian Cho, "Physicists Nightmare Scenario: The Higgs and Nothing Else," *Science* 315, no. 5819 (March 2007): 1657–1658, https://doi.org/10.1126/science.315.5819.1657.

be evidence for their mathematical schemes such as string theory,[634] or a model of larger particles called "supersymmetry." Supersymmetry proposes that every particle has a super partner, and a super partner for the Higgs boson would help explain its otherwise fine-tuned mass.[635]

Hopefully, the collider might even find evidence for microscopic black holes predicted to exist by string theory.[636]

To their dismay, the Large Hadron Collider confirmed the existence of the Higgs boson but nothing else. When asked why such a special field delivers mass, there is only one answer. "It is the result of an accident in the history of the universe in which a field froze in empty space in a certain way."[637]

Once Again, They Seem to Say, Mother Nature Caught a Lucky Break

Who buys that explanation? Just like the fractional charges of up and down quarks or the weak force creating deuterium and tritium by converting up quarks to down, the Higgs field reflects the mind of God. A superintelligent power commissioned the blueprints of creation before the beginning of time, including the Higgs field.

---

[634]   Sabine Hossenfelder, *Lost in Math: How Beauty Leads Physics Astray* (New York: Basic Books, 2018), 104.
[635]   Hossenfelder, *Lost in Math*, 11.
[636]   Bernard J. Carr and Steven B. Giddings, "Quantum Black Holes," *Scientific American*, April 1, 2007, https://www.scientificamerican.com/article/quantum-black-holes-2007-04/.
[637]   Krauss, *The Greatest Story Ever Told*, 304.

Understanding the Atom

The heat and light of the big bang challenges our imagination. But there was a feature to the light more astonishing than its temperature. The light could accomplish something that challenges human understanding. It could morph into matter, congealing into something called "fundamental particles" that could build a near-infinite universe filled with galaxies and stars. When high-energy photons collided in the newborn universe, they spun off a quark.[638] Quarks flowed with gluons in a quark-gluon plasma as the Higgs mechanism encoded quarks with mass.

By Way of Brief Review: Atoms Consist of Two Parts

1. A nucleus contains positively charged protons and neutrally charged neutrons. As fusion in the stars creates heavier elements, those elements add more protons and neutrons to the nucleus.

---

[638] Jonathan Allday, *Quarks, Leptons and the Big Bang*, 2nd ed. (Boca Raton: CRC Press, 2017).

2. Orbiting electrons hover above the nucleus and match the positive charge of protons with a negative charge of their own.

The History of Particle Physics: The Atom

In 440 BC, a Greek philosopher, Democritus, broke a stone in half and realized that the two halves had the same properties as the original stone. Carrying this concept to its logical conclusion, he realized it would be impossible to keep dividing the stone into an infinite number of smaller pieces. Democritus named the smallest possible piece of the stone "*atomos,*" a Greek word for "cannot be divided."[639]

Fast-forward to AD 1800—fascinated by the partial pressure of gases, chemist John Dalton used experimentation to develop the first scientific theory of the atom. He proposed that each chemical element is composed of atoms of a single, unique type—fundamental units that no technique can alter or destroy. All atoms of a given element are identical, but atoms of different elements have their own unique size and mass.[640]

Although Dalton was a modest Quaker, more than forty thousand people marched in his funeral procession in 1844.[641]

---

[639] Paul Cartledge, *Democritus* (New York: Routledge, 1999), 43.

[640] Elizabeth C. Patterson, *John Dalton and the Atomic Theory: The Biography of a Natural Philosopher* (New York: Doubleday, 1970).

[641] Kristine Krug, "Science Celebrates 'Father of Nanotech'," BBC News, October 10, 2003, http://news.bbc.co.uk/2/hi/science/nature/3178890.stm.

The Electron

Dalton believed his atoms were tiny, hard spheres with no component parts. Nothing could break them down into anything smaller. In 1897, J. J. Thomson proved him wrong.[642]

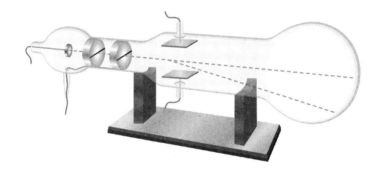

Thomson Cathode Ray Tube Experiment

J.J. Thomson discovered the electron in 1887, using a
cathode ray tube with positive and negative posts.

Using a cathode ray tube (much like the television sets of the 1950s), Thomson fired particles through a vacuum from a negative to a positive electrode. He discovered that atoms contained a tiny, negatively charged particle called the electron. This particle was shockingly small, containing 1/1800th of the mass of the atom. This little electron possessed a negative charge, but the overall atom was charge neutral. Believing that negatively charged electrons must float in a soup of positive protons, Thomson proposed a "plum pudding" model of the atom.

---

[642] E. A. Davis and Isabel Falconer, *J.J. Thompson and the Discovery of the Electron* (London: CRC Press, 2014).

Electrons are tiny fragments of energy that power our lights, smartphones, and balance the electric charge of every atom in our body. The brain is an electrical organ with brain waves that can be recorded in an EEG. Electrical pacemakers trigger the beats of our heart.[643]

Electrons are produced by the electromagnetic fields, and those fields have a force particle ("boson") called the photon. Who among us fails to recognize the wonders of the photon impacting our retina and delivering the details of the world "out there" to the neurons and synapses of the brain?

The Proton

Physicist Ernest Rutherford soon spoiled Thomson's pudding. Marie Curie, a French physicist, had discovered radium in 1898, and Rutherford soon realized that radium was emitting alpha particles—nuclei of the helium atom containing two positive protons and two neutrons.

In 1917, Rutherford placed a sample of Curie's radium inside a lead box with a tiny pinhole and fired a thin beam of alpha particles through a sheet of pure gold. If the plum pudding model was correct and negative and positive charges are distributed evenly throughout the atom, most of those alpha particles would pass through the gold undisturbed.[644]

Most of the particles behaved as expected, traveling a straight path through the foil to a detector. But the occasional particle ricocheted off the track by more than 90 degrees. Rutherford used the

---

[643] Arthur Firstenberg, *The Invisible Rainbow: A History of Electricity and Life* (White River Junction, VT: Chelsea Green Publishing, 2020).

[644] Richard Reeves, *A Force of Nature: The Frontier Genius of Ernest Rutherford* (New York: W. W. Norton and Company, 2007).

following analogy: "It was…as if you fired a 15-inch shell at a piece of tissue paper and it came back and hit you."[645]

Rutherford Quickly Came to Some Startling and Accurate Conclusions[646]

1. Since most particles passed quietly through the gold foil, the nucleus of the atom containing its positive charge and most of its mass must be extremely small, so small that alpha particles and atoms of gold rarely run into each other. Atoms, he properly concluded, are mostly empty space.

2. But the nucleus is also real: a few alpha particles were drastically redirected by a rare collision with a gold nucleus. The nucleus might be tiny, but when it makes a direct hit with a gold atom, it demonstrates its mass with a dramatic ricochet.

The modern model of the atom was complete. Positive protons live in the atom's tiny nucleus, while negatively charged electrons orbit overhead. The atom is 99.9 percent empty space. If we removed the empty space from our atoms, we would shrink to the size of a grain of salt![647]

The Neutron

Ernest Rutherford discovered the nucleus in 1911 and the proton in 1919. But Rutherford and his assistant, James Chadwick,

---

[645] Reeves, *A Force of Nature*, 74.

[646] John Rowland, *Ernest Rutherford, Atom Pioneer* (New York: Philosophical Library, 1957).

[647] Paul Sen, "Tiny Finding that Opened New Frontier," BBC News, July 25, 2007, http://news.bbc.co.uk/2/mobile/science/nature/6914175.stm.

realized that there must be something contributing additional mass to the nucleus of the atom without disturbing its electric charge. Helium, for instance, was known to have an atomic number of two (the number of protons) but a mass number of four. What provided the extra mass?

Chadwick proved the existence of a particle similar in mass to the proton but with no electric charge.[648] Science soon had a new particle, the neutron. Chadwick received a Nobel Prize, and scientists had a new tool to probe the atom. To explore atomic structure, physicists attempted to split the atom with a stream of neutrons.

Nuclear Fission

In the 1930s, an obvious question arose: could humanity extract energy from the atom? In a 1933 interview, Rutherford called such expectations "moonshine." Einstein compared particle bombardment with shooting in the dark at a widely scattered flock of birds. At the same time, Niels Bohr, a Danish Nobel laureate, agreed that the chances of extracting energy from the atom were remote.

Once Again, the World's Leading Scientists Were Wrong

Nazi Germany managed to split the atom in 1938 and made a startling observation: the products of their experiments *weighed less than the starting uranium*. They were converting mass into energy, according to Einstein's famous equation $E = mc^2$.[649]

---

[648] James Chadwick, "The Existence of a Neutron," *Proceedings of the Royal Society A* 136, no. 830 (June 1932): 692–708, https://doi.org/10.1098/rspa.1932.0112.

[649] Sam Kean, *The Bastard Brigade: The True Story of the Renegade Scientists and Spies Who Sabotaged the Nazi Atomic Bomb* (Boston: Little, Brown and Company, 2019).

Einstein reported the implications in a letter to President Franklin Roosevelt, and the Manhattan Project was soon underway.[650] Six years later, atom bombs destroyed Hiroshima and Nagasaki.

The Quark

Following the war, researchers continued to pursue high-energy physics, building more powerful colliders to split the atom into smaller pieces. President Dwight D. Eisenhower signed off on the budget for the two-mile-long SLAC National Accelerator Laboratory that could accelerate particles to an energy of fifty billion electron volts and focus this incredible energy into micron-sized targets to produce multiple particle collisions.[651]

In 1968, a series of electron-proton scattering experiments at SLAC revealed the first signs that protons and neutrons have point-like particles hiding inside.[652] Accelerators exposed the existence of the quark.

The Higgs Boson

Following completion of the standard model, one particle remained elusive. Scientists had proposed that Higgs bosons interact with electrons and quarks and encode those particles with mass. Without mass, the universe would have no matter or physical structure.

---

[650] Cynthia C. Kelly, ed., *The Manhattan Project: The Birth of the Atomic Bomb in the Words of Its Creators, Eyewitnesses, and Historians* (New York: Black Dog and Leventhal Publishers, 2020).

[651] John R. Rees, "The Stanford Linear Collider," *Scientific American* 261, no. 4 (October 1989): 58–65, https://www.jstor.org/stable/24987437.

[652] Frank Close, *The New Cosmic Onion: Quarks and the Nature of the Universe* (Boca Raton: CRC Press, 2006).

When the Higgs mechanism encodes particles with mass, they slow down, interact, and create protons, neutrons, and the nucleus of atoms.

The Higgs boson has a mass greater than 125 protons combined, so confirming the existence of the Higgs would require more energy than SLAC could provide. While SLAC produced the energy of fifty billion electron volts, to expose the Higgs, physicists would need to increase the power of collisions by two hundredfold.[653]

Scientists and the public alike marvel at the Large Hadron Collider (LHC) at the CERN laboratory near Geneva, Switzerland. Its basic features are mind-numbing: it provides a circular path 16.8 miles long, studded with 9,300 magnets, supercooled to -456.25 degrees Fahrenheit, enabling the collider to accelerate protons to 99.99 percent of the speed of light.[654] The LHC creates two beams of particles traveling in opposite directions. A succession of machines with increasingly higher energies accelerates the particles to target energy, then feeds them into the next machine in the chain. The apparatus consumes more power than a major city.

It took about a decade to construct the LHC at a total cost of about $4.75 billion. The cost for computing power for the LHC approaches $286 million annually, while electricity alone runs about $23.5 million per year. In daily experiments, the LHC generates enough data to fill twenty million tall filing cabinets.[655]

---

[653] James Gillies, *CERN and the Higgs Boson: The Global Quest for the Building Blocks of Reality* (London: Icon Books, 2018).

[654] Gillies, *CERN and the Higgs* Boson.

[655] Henry E. Brady, "The Challenge of Big Data and Data Science," *Annual Review of Political Science* 22 (May 2019): 297–323, https://doi.org/10.1146/annurev-polisci-090216-023229.

In 2012, collider experiments at CERN in Geneva, Switzerland, confirmed the existence of the Higgs boson.[656]

Experiments at the LHC continue to produce more energy, reaching maximum energies of fourteen trillion electron volts. Has the LHC produced another breakthrough? Sadly, the answer is "no." Particle physicists face a nightmare scenario: could the discovery of the Higgs represent the end of particle physics?

Questions Abound in Cosmology and Physics[657]

1. What is dark matter?
2. Is there a quantum theory of gravity?
3. If space-time is like a multilayered fishnet with intersecting loops, what ties those nodes or loops together?
4. Is there evidence for a holographic universe?
5. Is there evidence for a multiverse?

To answer these questions, particle physicists are lobbying for the International Linear Collider, capable of reaching one hundred trillion electron volts, seven times more potent than the LHC. Its construction would cost more than $10 billion. So far, no country has stepped forward to foot the bill.[658]

---

[656] Amir Aczel, "We (Apparently) Found the Higgs Boson. Now, Where the Heck Did It Come From?" *Discover Magazine*, last updated November 20, 2019, http://blogs.discovermagazine.com/crux/2012/07/09/we-apparently-found-the-Higgs-boson-now-where-the-heck-did-it-come-from/.

[657] Sabine Hossenfelder, *Lost in Math: How Beauty Leads Physics Astray* (New York: Basic Books, 2018), 12.

[658] Toni Feder, "CERN Considers a 100 TEV Circular Hadron Collider," Physics Today, February 5, 2019, https://doi.org/10.1063/PT.6.2.20190205a.

A Double Dose of Carbon for Life

We are 60 percent water, or H2O. The next most abundant element in our body is carbon at 18 percent.[659] Life as we know it requires a generous supply of carbon. Carbon-based molecules such as DNA and proteins build our cells and tissues. God used quantum superposition and entanglement to double the production of carbon in the stars.

We can imagine how particles take a quantum leap by visualizing two parachutists photographed in free fall, holding hands after jumping out of their airplane.

With a stretch of your imagination, believe these parachutists have taken a quantum leap—up from the ground. Imagine that these two parachutists represent two helium nuclei that have jumped to the same energy level in the center of a star.

These helium nuclei have become entangled and will soon fuse to become an atom of beryllium (helium-4 + helium-4 = beryllium-8).

---

[659] Michael Schirber, "The Chemistry of Life: The Human Body," Live Science, April 16, 2009, https://www.livescience.com/3505-chemistry-life-human-body.html.

Review the table of the elements organized by their increasing size. Hydrogen, consisting of a single proton, sits at the top left of the periodic table. During nuclear fusion in the sun, hydrogen becomes helium, an atom with two protons and two neutrons, sitting at the top of the table on the right. The next row consists of heavier elements, lithium, beryllium, boron, and carbon. Fusion later proceeds to nitrogen, oxygen, and fluorine.

Interloping Carbon

Before those two helium nuclei can fuse into beryllium, a third helium leaps into the process. In our imaginary model, a third parachutist moves into a three-way handhold. Why should that matter? Because carbon-12 (the fusion product of the three helium nuclei) takes the exact same quantum leap as beryllium-8.[660] Unstable beryllium disappears, and the output of carbon is doubled.

Fred Hoyle, a nuclear physicist in the 1960s, realized that the stars might fail to create sufficient carbon for life. As a result, he predicted that carbon would have the same quantum resonance as beryllium in the stars. Analysis of resonance levels for carbon soon confirmed his prediction.[661] Carbon and beryllium take precisely the same quantum leap, and carbon is the victor in the process. Is it surprising that carbon was a priority product of the stars?

The Effort to Build a Quantum Computer

Any discussion of quantum mechanics should mention the ongoing effort of IBM, Google, and other major companies to build a

---

[660] Paul Halpern, *Flashes of Creation: George Gamow, Fred Hoyle, and the Great Big Bang Debate* (New York: Basic Books, 2021), 15.
[661] Jane Gregory, *Fred Hoyle's Universe* (Oxford: Oxford University Press, 2005).

quantum computer.[662] Standard computers depend on the flow of electrons through silicon transistors. The basic unit of memory is a bit. The bit has just two potential outputs, a zero or one. How would a quantum computer differ from our average laptop?

In a quantum computer, the basic unit of memory is a quantum bit or qubit. Qubits exploit the quantum magic of a fundamental particle, such as the spin of an electron or the polarization of a photon. Thanks to the power of superposition and entanglement, these systems can be in many different arrangements at once. Quantum computer bits are both zero and one, on and off in the same instant, retaining all outcomes until reaching a final answer.[663] Quantum computers would provide the information processing power required to solve problems too difficult for classical computers.

Instead of analyzing zeros and ones sequentially, two qubits in superposition can represent four scenarios at the same time. Should you ask a classical computer for the best path out of a maze, the traditional computer would try every path one at a time. A quantum computer would try all routes at once, "holding uncertainty in its head" until reaching a definitive answer.[664]

Quantum computers require special algorithms, or step-by-step instructions, to tackle complex data. Peter Schor developed the first quantum algorithm in 1994 to factor complex numbers, a capability with enormous implications for artificial intelligence and cyber security.[665]

---

[662] Amit Katwala, "Quantum Computing and Quantum Supremacy, Explained," *Wired*, May 5, 2020, www.wired.co.uk/article/quantum-computing-explained.

[663] Chris Bernhardt, *Quantum Computing for Everyone* (Cambridge, MA: MIT Press, 2020).

[664] Katwala, "Quantum Computing."

[665] P. W. Shor, "Algorithms for Quantum Computation: Discrete Logarithms and Factoring," *Proceedings 35th Annual Symposium on Foundations of Computer Science*, November 20–22, 1994, 124–134.

The quantum capabilities of the particles of creation were essential to God's creative plan. Imagine the constant entanglement of waveforms as nuclear fusion lights up a star. Imagine how many wave function mergers and acquisitions happen as fundamental particles weave the miracle of life. Imagine the information processing power, should an entire universe function as a quantum computer.

The quantum wave function is a fundamental blueprint of God's creative plan.

The Fabric of Space-Time

Einstein's Nobel

Einstein did not win the Nobel Prize for his theory of general relativity. But he did win the Nobel for his work with the photoelectric effect. Einstein provided the scientific foundation for the garage door opener and the solar panel.[666]

Following his miracle year and the publication of four astounding articles in 1905, Einstein spent the next decade attempting to broaden his special theory of relativity into a general theory. His goal: explain gravity as the distortion of space-time's geometry.

His breakthrough arrived in the form of Riemannian geometry. Developed in 1854, Riemannian geometry dealt with the geometry of curved surfaces.[667] Space-time curved in the presence of energy or mass, and that curvature created the phenomenon of gravity.

As a result of this space-time curvature, every mass-bearing object in the universe attracts every other object with mass. It may be more accurate to say that objects with mass slide down space-time's curvature and fall toward each other. We fall toward

---

[666] Walter Isaacson, *Einstein: His Life and Universe* (New York: Simon and Schuster, 2007), 96.

[667] Albert Einstein, *Relativity: The Special and the General Theory - 100th Anniversary Edition* (Princeton: Princeton University Press, 2015).

Earth. The moon falls toward Earth as well but has enough angular momentum to remain in a stable Earthbound orbit.

What More Can We Learn about Space-Time

In addition to viewing space-time as multilayered bubble wrap, we can also compare it to a four-dimensional fishnet. Tiny knots tie each node of a fishnet together. We have arrived at the deep mystery of space-time. What produces the nodes of space-time, weaving the space-time fabric?

String Theory

For decades, string theorists thought they had the answer.[668] Our universe, they proposed, has ten or more hidden higher dimensions. All but four of those dimensions curl up together to create a harmonica-like musical instrument called a "Calabi-Yau manifold." Those harmonicas produced by higher dimensions become the node of space-time and play the music of the strings.[669]

As strings travel through these tiny musical instruments, they vibrate and spin in diverse ways, producing the particles of the standard model. According to string theory, these tiny musical manifolds also represent the nodes of space-time.

---

[668] Volker Schomerus, *A Primer on String Theory* (Cambridge: Cambridge University Press, 2017).

[669] Lisa Randall, *Warped Passages: Unraveling the Mystery of the Universe's Hidden Dimensions* (New York: Ecco, 2006), 41.

Hopes for a Theory of Everything

String theorists hoped their models might pave the way to a theory of everything, uniting Einstein's relativity with quantum mechanics. They also believed that they had found a leading candidate for a theory of a multiverse. Those ten hidden higher dimensions of string theory could collapse in infinite ways, each collapse producing a unique universe with its own set of fields and forces.[670]

Supersymmetry Dreams

String theorists also support a theoretical model of the universe referred to as supersymmetry.[671] Each particle in the standard model would have a larger supersymmetric partner, creating a set of "super" particles that complete the standard model, like the top half of a clam shell. Supersymmetry, if true, would explain the fine-tuning of the Higgs mass and might also reveal the source of dark matter.

String theorists produced untold numbers of publications and shelves full of books,[672] but they faced an insurmountable problem. There was no effective way to test their theories. Their best hope for laboratory confirmation, making string theory real science, would come from the Large Hadron Collider (LHC), the same apparatus that proved the existence of the Higgs boson.

Physicists prayed that the LHC would provide evidence for supersymmetric particles. A supersymmetric partner for the Higgs

---

[670] Lee Smolen, *The Trouble with Physics: The Rise of String Theory, the Fall of a Science, and What Comes Next* (Boston: Houghton Mifflin Harcourt, 2006).

[671] Sabine Hossenfelder, *Lost in Math: How Beauty Leads Physics Astray* (New York: Basic Books, 2018).

[672] Brian Greene, *The Fabric of the Cosmos: Space, Time, and the Texture of Reality* (New York: Vintage Books, 2005).

boson would explain its perfect mass. The LHC might find evidence for microscopic black holes, reflecting the reality of hidden dimensions.

Engineers have pushed the LHC to the limits of its power, but no new data has emerged.[673] String theory is steadily losing favor, and other more promising theories are taking its place.[674]

New Theories About the Nature of Space-Time

Scientists continue to grapple with the mysteries of space-time. One leading theory: the nodes of space-time are quantum bits, or "qubits." We may live in a quantum computer.[675] An even more startling possibility: our universe may be a hologram.[676]

---

[673] Adrian Cho, "Physicists Nightmare Scenario: The Higgs and Nothing Else," *Science* 315, no. 5819 (March 2007): 1657–1658, https://doi.org/10.1126/science.315.5819.1657.

[674] Smolen, *The Trouble with Physics*, 161.

[675] Seth Lloyd, *Programming the Universe: A Quantum Computer Scientist Takes on the Cosmos* (New York: Vintage Books, 2007), 70.

[676] Anil Ananthaswamy, "Is Our Universe a Hologram? Physicists Debate Famous Idea on its 25th Anniversary," *Scientific American*, November 30, 2022, https://www.scientificamerican.com/article/is-our-universe-a-hologram-physicists-debate-famous-idea-on-its-25th-anniversary/.

Black hole with an event horizon. The force of gravity is so great within a black hole, not even light can escape.

String theory produced unexpected and powerful dividends when applied to the analysis of black holes, those mysterious objects produced by the collapse of massive stars. Inside a black hole, gravity is so strong that nothing, not even light, can escape its embrace. Matter consumed by a black hole is squeezed into a tiny space, but the black hole keeps a record of its contents. Information about the contents of the black hole's bulk is displayed on its surface, a disc-like, two-dimensional circle called an "event horizon."[677]

This analysis of black holes gave rise to the "holographic principle": the description of a volume of space can be encoded on a lower-dimensional boundary. In short order, this principle was applied to the entire universe. As noted in chapter 5, Leonard

---

[677] Stephen Hawking, *Black Holes: The Reith Lectures* (New York: Penguin Random House, 2001).

Susskind, professor of theoretical physics at Stanford University and founding director of the Stanford Institute for Theoretical Physics, made a bold proposal: "The three-dimensional world of ordinary experience—the universe filled with galaxies, stars, planets, houses, boulders, and people—is a hologram, an image of reality coded on a distant two-dimensional surface."[678]

In 1997, Juan Maldacena of the Institute for Advanced Study supported this conclusion as he studied an imaginary model of the universe called anti-de Sitter space (AdS).[679]

The imaginary AdS universe has five dimensions and includes gravitational force. But Maldacena made a second observation: a lower-dimensional, gravity-free quantum field described by conformal field theory (CFT), describes the same universe. Mathematics in the anti-de Sitter space and mathematics in the conformal field describe the same physical system.

Maldacena reached a logical conclusion: our universe and its space-time may not be fundamental. Space-time might be a holographic projection of a lower-dimensional reality—like fog rising from the surface of a pond.

During the subsequent twenty-five years since the development of these model universes, physicists have used AdS/CFT duality to address fundamental questions about our universe.

1. Stephen Hawking demonstrated that black holes could emit radiation and slowly shrink away. When a black hole finally disappears, what happens to the information stored

---

[678] Leonard Susskind, *The Black Hole War: My Battle with Stephen Hawking to Make the World Safe for Quantum Mechanics* (New York: Little, Brown and Company, 2008), 410.

[679] Juan Maldacena, "The Large N limit of Superconformal Field Theories and Supergravity," *International Journal of Theoretical Physics* 38 (April 1999): 1113–1133, https://doi.org/10.1023/A:1026654312961.

on their event horizon? Has it been lost, which would violate a fundamental rule of quantum mechanics? Or has it been preserved on the conformal field?

2. Can the emergence of space-time through entanglement with a conformal field explain the rapid expansion of the newborn universe, a phenomenon known as inflation, that made the universe flat and isotropic (looking the same in all directions)?

3. Most importantly, can further refinement of AdS/CFT duality lead to a quantum theory of gravity, merging Einstein's theory of relativity with the principles of quantum mechanics?

4. Finally, the reality of a holographic universe should give us pause when dismissing the possibilities of miracles. In an AdS/CFT universe, could God use His mathematical powers to manipulate the conformal field, perhaps to cure a leper, turn water into wine, heal a withered hand, or resurrect His son! Occupants of the holographic bulk (theoretically, us) would be oblivious to the source of miraculous events. Would such a quantum field give God access to His universe without violating His physical laws?

Building a Star on Earth

If humankind could transport the power of the sun to the surface of Earth, creating an Earthbound fusion reactor, the world would have an eternal supply of "clean energy." Nuclear fusion reactors would decarbonize our world. But an Earthbound fusion reactor has long been a dream that is "thirty years away…and always will be."[680]

The initial challenge is obvious—the diameter of the sun is more than one hundred times greater than our planet, and its mass exceeds Earth's mass by more than three hundred thousand fold.[681] The force of gravity at the surface of the sun is twenty-nine times stronger than gravity on Earth. How can we generate sufficient a squeezing force on Earth to fuse deuterium and tritium into helium?

Efforts to Maximize the Squeeze

Existing nuclear power plants depend on nuclear fission, harvesting the release of energy when heavy atoms, such as uranium, undergo decay. Nuclear meltdowns at Three Mile Island, Chernobyl, and Fukushima have demonstrated the dangers associated with nuclear fission.

---

[680] Arthur Turrell, *The Star Builders: Nuclear Fusion and the Race to Power the Planet* (New York: Scribner, 2021), 11.
[681] Turrell, *The Star Builders*, 77.

In contrast, fusion merges very light atomic nuclei, releasing .07 percent of the starting mass as pure energy multiplied by the speed of light squared. Nuclear fusion would single-handedly solve the problem of carbon-based global warming. But fusion can only happen at extreme levels of temperature and pressure.

Modern industry continues to chase the dream. The Fusion Industry Association in Washington DC, reports that more than thirty commercial companies are pursuing a fusion reactor.[682]

Efforts so far have involved heating deuterium and tritium together until they form plasma, a fluid state of matter.[683] These isotopes of hydrogen enter a plasma state more readily than normal hydrogen, but only fuse at temperatures exceeding one hundred million degrees kelvin. Efforts to stabilize plasma with magnetic fields at these extreme temperatures have been challenging, and brief moments of fusion have produced disappointing quantities of energy.

The stated target by 2035 (the proverbial multiyear dream away) is to produce five hundred milliwatts of power from an input of fifty milliwatts.[684] This output is comparable to the energy production of a single coal-fired power plant, but nuclear fusion would rapidly become the preferred source of energy around the world.

682   Philip Ball, "The Race to Fusion Energy," *Nature* 599 (November 2021): 362–366, https://www.nature.com/immersive/d41586-021-03401-w/index.html.
683   Turrell, *The Star Builders*, 63.
684   Peter Dunn, "Can the U.S. Switch on a Nuclear Fusion Plant by 2035?" Energypost.eu, March 8, 2021, https://energypost.eu/can-the-u-s-switch-on-a-nuclear-fusion-plant-by-2035/.

A device called a Tokamak partially recreates the heat and pressure driving nuclear fusion in the stars. Most scientists predict that humankind will build "a star on Earth" in the next few decades, giving us free and abundant energy.

The most common approach to controlling plasma involves placing the reaction in a container called a "tokamak" and using powerful superconducting magnets to contain the plasma.[685] Companies are experimenting with different forms of tokamak, as well as with new designs for more powerful magnets. They are also using "machine learning," or artificial intelligence, to track turbulence at extreme temperatures and pressures.

One company, First Light Fusion, a private spinoff of Oxford University, is abandoning the need for magnetic fields. This fusion laboratory compresses plasma with a "shockwave" using an electromagnetic projectile gun to fire a material (identity not yet revealed) into a target of the hydrogen isotopes at speeds exceeding fifty

---

[685] H. Han et al., "A Sustained High-Temperature Fusion Plasma Regime Facilitated by Fast Ions," *Nature* 609 (2022): 269–275, https://doi.org/10.1038/s41586-022-05008-1.

kilometers per second.[686] Other companies are experimenting with piston compression and high-speed centrifugation and the use of abundant "fast ions" to stabilize plasma turbulence. The universal goal: make small fusion reactors compatible with existing energy grids by the end of the current decade.

Precious Tritium

Another challenge faces the effort to produce nuclear fusion on Earth. Deuterium can be produced in abundance from sea water. Tritium is another matter. The hydrogen isotope called tritium undergoes radioactive decay, leaving scant quantities of tritium on Earth.[687] The techniques to produce tritium are expensive, producing small amounts of the hydrogen isotope at costs exceeding $30,000 *per gram*. That represents a major problem for reactor designs. Each one will require five hundred kilograms of tritium per year.

Will humankind overcome these challenges and bring solar-like nuclear fusion to reactors on planet Earth? There is only one reason to believe it might happen. God created men and women in His image, and He encourages our efforts to harness the created order. A recent report from the Lawrence Livermore National Laboratory and its National Ignition Facility described a major step forward. An experiment approached "ignition" as the world's most

---

[686] Ed Brown, "Gigantic,70-Foot Nuclear Fusion Gun Could Change the World," *Newsweek*, October 28, 2022, https://www.newsweek.com/first-light-fusion-nuclear-gun-uk-1754934.

[687] Richard J. Pearson, Armando B. Antoniazzi, and William J. Nuttall, "Tritium Supply and Use: A Key Issue for the Development of Nuclear Fusion Energy," *Fusion Engineering and Design* 136, part B (November 2018): 1140–1148, https://doi.org/10.1016/j.fusengdes.2018.04.090.

energetic laser bombarded a peppercorn-size fuel capsule.[688] The reaction produced 70 percent of the input energy, more than twice of any prior energy return.

Even more recently (December 2022), the US Energy Department announced another breakthrough. Lasers at the Lawrence Livermore Laboratory heated and compressed a capsule of hydrogen to record levels of temperature and pressure, triggering a fusion reaction that produced 50 percent more energy than the energy input. For the first time in human history, nuclear fusion on Earth delivered a net gain of energy.[689] But don't expect fusion reactors anytime soon. It will take decades of more research to make the model cost competitive. Delivering the power of the sun to the surface of Earth is still twenty to thirty years away.

Meanwhile, a nuclear reactor called the sun continues to produce free and abundant tritium as it delivers perfect heat and light to a rare and privileged Earth.

---

[688] Daniel Clery, "With Explosive New Result, Laser-Powered Fusion Effort Nears 'Ignition'," *Science*, August 17, 2021, https://www.science.org/content/article/explosive-new-result-laser-powered-fusion-effort-nears-ignition.

[689] Steven E. Koonin and Robert L. Powell, "How Fusion Works and Why It's a Breakthrough," *Wall Street Journal*, December 14, 2022, https://www.wsj.com/articles/how-fusion-works-and-why-its-a-breakthrough-national-ignition-facility-laser-beam-laboratory-science-nuclear-weapon-11671052081.

Electrifying Life

Electricity lights our homes and streets, warms and cools our air, and quietly animates our computers and smartphones. But this miracle of modern life has only been broadly available for less than a century, and electricity remains beyond the reach of many citizens in the underdeveloped world.

People have been aware of electricity since ancient times, but Benjamin Franklin's kite experiment demonstrated how poorly it was understood.[690] Attaching a metal key to a kite, Franklin demonstrated that lightning is a form of static electricity. In response to his potentially fatal kite experiment, he invented the lightning rod to guide lightning bolts safely to the ground.[691]

The British scientist Michael Faraday first tamed the electromagnetic force by inducing an electric current with magnets inside coils of copper wire.[692] A few decades later, James Clerk Maxwell supported this discovery with a series of beautiful equations.[693]

---

[690] Walter Isaacson, *Benjamin Franklin: An American Life* (New York: Simon and Schuster, 2003), 140.
[691] Isaacson, *Benjamin Franklin*, 143.
[692] Jill Jonnes, *Empires of Light: Edison, Tesla, Westinghouse, and the Race to Electrify the World* (New York: Random House, 2004), 53.
[693] Andrew Zangwill, *Modern Electrodynamics* (Cambridge: Cambridge University Press, 2012).

Thomas Edison created the light bulb and delivered electricity
to buildings in New York city in the form of direct (DC) current.
Tesla's delivery of electricity as alternating current tamed
by transformers successfully electrified the nation.

Most of us credit Thomas Edison with turning on our modern
lights. Through tireless work in the 1870s, he developed the first
incandescent light bulb, leading a German historian to declare:
*"Fire had been discovered for the second time...mankind had been
delivered again from the curse of the night."*[694]

---

[694] Robert L. Bradley, *Edison to Enron: Energy Markets and Political Strategies*
(Hoboken: Wiley-Scrivener, 2011), 30.

Early inventions spurred the development of the field. The telegraph gave way to the telephone as lightbulbs, televisions, and kitchen appliances created the modern world. Today, the internet and cloud data storage demand large quantities of reliable electricity.

General Electric

Edison used small generators and copper wires to deliver electric lighting to a small number of New York homes, then launched the first business dedicated to electric power with a central power plant in Manhattan. That company is known today as General Electric.[695] But Edison's model for large-scale delivery of electric power was flawed. His electricity flowed as a direct current. Over long distances, direct current overheats cables and loses unacceptably large quantities of energy.[696]

Nikola Tesla, an Edison employee, proposed an alternative model.[697] Electricity should be distributed as an alternating current, allowing high voltage to be distributed over great distances with minimal energy waste.[698] The only downside: alternating current involves dangerous voltage and the threat of fatal shock.

Transformers Saved the Day

A discovery called the transformer saved the day.[699] Transformers reduce the voltage of alternating current before distribution to factories or homes.

---

[695] Edmund Morris, *Edison* (New York: Random House, 2020), 94.
[696] Jonnes, *Empires of Light*, 10.
[697] Jonnes, *Empires of Light*, 122.
[698] Jonnes, *Empires of Light*, 10.
[699] Morris, *Edison*, 471.

Westinghouse

The advantages of the Tesla system prevailed in this "war of the currents." Westinghouse bought Tesla's patents and won the bid to build a hydroelectric power station at Niagara Falls.[700] The use of direct current rapidly declined as alternating current electrified the nation.

Fast-forward to the modern day. We generate electricity in bulk using coal-fired power plants, hydroelectric sources, natural gas, and nuclear reactors. Spurred by the risks of climate change, scientists are attempting to decarbonize electrical generation with solar panels and wind turbines. A nuclear fusion reactor would provide the ultimate solution, but as discussed, fusion power plants are likely still decades away.

Five Hundred Million People Still Live in the Dark

Thirty percent of our fossil fuel consumption is devoted to the generation of electric power, reflecting its silent efficiency and adjustable flow. Despite its embrace by developing countries around the world, over five hundred million people continue to live in the dark with no access to electric lights.[701]

---

[700] Morris, *Edison*, 327.
[701] Hannah Ritchie, "The Number of People Without Electricity More Than Halved Over the Last Twenty Years," Our World in Data, November 30, 2021, https://ourworldindata.org/number-without-electricity.

Discovering the Universe

Earth has been called the "privileged planet," located on a spiral arm of the Milky Way with an unobstructed view of the universe that encourages study and observation.[702] Scientists have exploited this "privilege," building ever more sophisticated instruments with which to view our cosmos and its ancient past.

Space Telescopes

The atmosphere of Earth partially obscures our deep-space view. In 1990, the Hubble telescope was launched into outer space to provide an atmosphere-free view of the cosmos.[703] A camera aboard Hubble has taken wide field and high-resolution images of galaxies, planets, and extra-galactic objects with resolutions tenfold greater than the largest Earthbound observatory. In 1993, a NASA space shuttle corrected deformities in the telescope's optics, allowing the Hubble to produce spectacular photographs of our cosmos. This telescope has revolutionized the field of astronomy.

---

[702] Guillermo Gonzalez and Jay W. Richards, *The Privileged Planet: How Our Place in the Cosmos Is Designed for Discovery* (Washington DC: Regnery Publishing, 2004).

[703] Jim Bell, *Hubble Legacy: 30 Years of Discoveries and Images* (New York: Sterling 2020).

Hubble photographs allowed the most accurate calculation of the Hubble constant, which we use to determine the rate of cosmic expansion.[704] It has photographed young stars giving birth to planetary systems. Photographs of more than one thousand galaxies have revealed the secrets of galactic evolution over the course of deep time.

James Webb Space Telescope

As Hubble has neared the end of its useful life, NASA launched the James Webb Space Telescope to continue space exploration.[705] By gathering infrared light, the telescope will capture photons that have traveled through space for over thirteen billion years, beginning their journey in the first ancient stars. James Webb Space Telescope images will test the latest theories of galaxy formation and universe expansion

Cosmic Background

Scientists have also launched satellites into space to detect and analyze the faint flow from the big bang called the cosmic microwave background, produced by the first release of photons as the universe cooled and suddenly became transparent.[706] Beginning with NASA's Cosmic Background Explorer in the 1990s and successor projects called the Wilkinson Microwave Anisotropy Probe, scientists have analyzed the cosmic microwave background in detail

---

[704] Gale E. Christianson, *Edwin Hubble: Mariner of the Nebulae* (Boca Raton, Florida: CRC Press, 1997), 194.

[705] Clara Moskowitz, "How JWST Is Changing Our View of the Universe," *Scientific American*, December 1, 2022, https://www.scientificamerican.com/article/how-jwst-is-changing-our-view-of-the-universe/.

[706] "COBE," NASA, June 16, 2023, https://science.nasa.gov/mission/cobe/.

Humankind has analyzed the universe with diverse equipment. This satellite analyzes the Cosmic Background Radiation (CBR) produced by the release of free photons in the early universe. Studies of the CBR have confirmed the Big Bang and cosmic expansion.

Studies of this ancient light have confirmed that the universe is flat.[707] Analysis of the cosmic background radiation has also supported the big bang model of creation.[708]

Gravitational Waves

In 1999, scientists created the Laser Interferometer Gravitational-Wave Observatory, usually referred to as LIGO, an observatory and large-scale physics experiment, in hopes of proving the existence of gravitational waves flowing across the universe after a major shock to the space-time fabric.[709] In 2017, a collision of two

---

[707] P. de Bernardis et al., "A Flat Universe from High-Resolution Maps of the Cosmic Microwave Background Radiation," *Nature* 404 (2000): 955–959, https://doi.org/10.1038/35010035.

[708] "Our Universe", Universe 101, NASA, last updated February 20, 2024, https://wmap.gsfc.nasa.gov/universe/universe.html.

[709] Brian Clegg, *Gravitational Waves: How Einstein's Spacetime Ripples Reveal the Secrets of the Universe* (London: Icon Books, 2018), 104.

neutron stars sent gravitational waves rippling across the heavens. LIGO detected the waves and reached another startling conclusion: these massive stars make some of the heaviest elements, including gold, silver, and uranium.[710]

Detecting Incoming Threats

While comets provide spectacular viewing, human beings should pay closer attention to asteroids, which are the rocky cousins of comets. Over eight hundred thousand of these missiles are flying through our galaxy in unpredictable orbits. Sixty-five million years ago, an asteroid collided with Earth on the Yucatán Peninsula in Mexico, creating a crater ninety miles wide. The result was a global firestorm, followed by a worldwide cold snap, rebounded by global warming. The event extinguished the dinosaurs and allowed mammals to take center stage.[711]

In 1908, an asteroid exploded over the Siberian Forest, killing untold numbers of reindeer and eighty million trees.[712] Scientists estimate that asteroids have struck the Earth 350 times in the past ten thousand years, and computer simulations suggest that asteroid impact could kill millions of people in the next ten thousand years.[713]

NASA has responded to this risk with a mission called the Double Asteroid Redirection Test, or DART. Should we detect an asteroid that could theoretically collide with Earth, NASA hopes to

---

[710] Sid Perkins, "Neutron Star Mergers May Create Much of the Universe's Gol," *Science*, March 20, 2018, https://www.science.org/content/article/neutron-star-mergers-may-create-much-universe-s-gold.

[711] Riley Black, *The Last Days of the Dinosaurs: An Asteroid, Extinction, and the Beginning of Our World* (New York: St. Martin's Press, 2022).

[712] Peter Jenniskens et al., "Tunguska Eyewitness Accounts, Injuries, and Casualties," *Icarus* 327 (July 2019): 4–18, https://doi.org/10.1016/j.icarus.2019.01.001.

[713] D. Perna et al., "A Global Response Roadmap to the Asteroid Impact Threat: The NEOshield Perspective," *Science* 118 (December 2015): 311–317, https://doi.org/10.1016/j.pss.2015.07.006.

slam a satellite into that asteroid with enough impact to bump it off course.[714] In a real-world test, the first DART mission scored a bullseye and altered a space rock's orbit.[715]

The Following are a Few Facts About Our Universe:

The universe is larger than we can comprehend and has no edges.[716]

The collapse of massive stars can create black holes with a gravitational field so intense that light is trapped in its grip. Peppered throughout the universe, these "stellar mass" black holes are ten to twenty-four times as massive as the sun.

On the other end of the size spectrum, "supermassive" black holes are millions, if not billions of times as massive as the sun. Astronomers believe that supermassive black holes lie at the center of all large galaxies, including our own Milky Way.[717]

Quasars are among the brightest objects in the universe. They are formed as material falls into the accretion disc (made of rapidly rotating gas) around a supermassive black hole. The incoming matter gets heated by friction and gravity and emits enormous amounts of energy.[718]

---

[714] "Double Asteroid Redirection Test (DART)," NASA, last updated June 26, 2024, https://science.nasa.gov/planetary-defense-dart/.

[715] Amit Malewar, "Impact Success! NASA's DART Successfully Impacted its Asteroid Target," Tech Explorist, September 27, 2022, https://www.techexplorist.com/impact-success-nasa-dart-successfully-impacted-asteroid-target/53965/.

[716] J. Richard Gott III et al., "A Map of the Universe," *The Astrophysical Journal* 624 (May 2005): 463–484, https://doi.org/10.1086/428890.

[717] A. M. Ghez et al., "High Proper-Motion Stars in the Vicinity of Sagittarius A*: Evidence for a Supermassive Black Hole at the Center of Our Galaxy," *The Astrophysical Journal* 509 (December 1998): 678–686, https://doi.org/10.1086/306528,

[718] Xue-Bing Wu et al., "An Ultra-Luminous Quasar with a Twelve-Billion-Solar-Mass Black Hole at Redshift 6.30," *Nature* 518 (February 2015): 512–515, https://doi.org/10.1038/nature14241.

Searching for Life Among the Stars

Scientists have been searching for life in the universe for decades. The SETI@home project, based at UC Berkeley, searched for signs of extraterrestrial intelligence using distributed computing on the internet.[719] They enlisted 180,000 active participants who volunteered a total of over 290,000 personal computers to analyze radio waves from outer space. SETI@home failed to detect evidence that ET exists.

SETI 2.0

NASA has discontinued SETI@home, replacing it with SETI 2.0.[720] In addition to searching for radio signals, SETI 2.0 will be looking for other signs of alien intelligence, including chemical changes in a planet's atmosphere.

---

[719] Daniel Uberhaus, SETI@Home is Over. But the Search for Alien Life Continues, *Wired*, March 3, 2020, https://www.wired.com/story/setihome-is-over-but-the-search-for-alien-life-continues/.

[720] Paul Scott Anderson, "Will SETI 2.0 Lead to a Discovery of Intelligent Aliens?" EarthSky April 14, 2020, https://earthsky.org/space/seti-breakthrough-listen-new-technologies-and-strategies/.

Icing on the Copernican Revolution

The detection of intelligent life on another planet would be icing on the cake for the sixteenth- century Copernican revolution, when scientists declared:

1. Earth was not the center of the universe, and
2. Human beings are not that special.

The longer the search goes on in silence, we may be tempted to reach a different conclusion:

Earth may the center of God's focus, if not the center of His universe.

More Than "Rare"

Other scientists have taken this argument to a higher level, arguing that Earth is more than just "rare." God has placed Earth in a perfect location to see and explore the universe. Dubbing Earth the "privileged planet," these observers note that our position in our galaxy, the Milky Way, gives us the clearest possible view of the universe.[721] Our atmosphere is clear. We are located on a spiral arm of our galaxy, not in the busy center of the Milky Way. Gravity stabilizes our rotation, and the sun and our moon are just the right size and distance from the Earth.

More Than "Luck"

Were we lucky in this regard? Guillermo Gonzalez, research scientist at the University of Alabama-Huntsville, and Jay Richard,

---

[721] Guillermo Gonzalez and Jay W. Richards, *The Privileged Planet: How Our Place in the Cosmos Is Designed for Discovery* (Washington DC: Regnery Publishing, 2004).

research fellow at the Heritage Foundation, believe this was more than mere luck: "We don't think this is merely coincidental. It cries out for another explanation, an explanation that suggests there's more to the cosmos than we have been willing to entertain or even imagine."[722]

Reading between the lines, these experts wonder if a superintelligence shaped the Earth, giving its inhabitants the best possible view of creation. God, it seems, wants us to understand and celebrate sacred science.

---

[722] Gonzalez and Richards, *The Privileged Planet*, xi.

Earth is Round

In ancient times, some primitive cultures may have believed the Earth was flat. Around 500 BC, Pythagoras looked at the spherical moon and proposed another model: the Earth is round, like the moon.[723] A century later, Aristotle observed that ships disappear hull first when they sail over the horizon. He also noted that the Earth casts a round shadow on the moon during a lunar eclipse.[724] Another ancient genius joined the round Earth club.

The First Measurement of Earth's Circumference

In Alexandria, Egypt, in 294 BC, the Greek mathematician Eratosthenes proposed to make a map of the world. As a first step, he decided to calculate its diameter. He recalled stories about a well in an Egyptian city known today as Aswan, with a bottom that remained shadow-free at noon on the summer solstice. Eratosthenes realized that the sun, on that day and time, must be directly over the well.

Placing a stick in the ground at his home in Alexandria, he measured the angle of the shadow cast by the sun on the summer

---

[723] "Starchild Question of the Month for February 2003," StarChild, https://starchild.gsfc.nasa.gov/docs/StarChild/questions/question54.html.

[724] "Starchild Question of the Month for February 2003,"

solstice: it was 7.2 degrees or 1/50th of a 360-degree circle. He then hired surveyors to count the steps between his home and the no-shadow well. The result? His stick and the well were 525 miles apart. Eratosthenes then calculated the circumference of the Earth by multiplying 50 by 525 miles, arriving at a circumference of 25,000 miles.[725]

The circumference of the Earth as measured by GPS today is 24,900 miles.[726]

Henry the Navigator

In the early fifteenth century, Prince Henry of Portugal, nicknamed "Henry the Navigator," financed many voyages of discovery around Africa and Asia.[727] Christopher Columbus learned to sail on Henry's journeys of exploration. But when he sought financing for a westward trip of his own—a journey to define a "shortcut" to the "Indies"—Prince Henry immediately realized Columbus was utilizing overly optimistic maps. Sailors had previously reached the Far East hugging the coast of Africa, and Henry knew the distance sailing westward would be much further.

Columbus: Unrealistic Dreamer

Columbus estimated the distance between the Canary Islands and Japan to be about 2,300 miles. Using Eratosthenes's estimate of the Earth's circumference, Henry knew the distance between the

[725] Duane W. Roller, *Eratosthenes' Geography* (Princeton: Princeton University Press, 2010).
[726] Robert Lea "How Big is Earth?" Space.com, July 7, 2021, https://www.space.com/17638-how-big-is-earth.html.
[727] Peter Russell, *Prince Henry the Navigator: A Life* (New Haven: Yale University Press, 2001).

Canary Islands and Japan was closer to 12,000 miles. He dismissed Columbus as an unrealistic dreamer.[728]

Columbus presented his optimistic projections to King Ferdinand and Queen Isabella of Spain. Hungry for gold and new converts to the Catholic faith, the Spaniards took the bait.

Fortunately for Columbus, the Caribbean archipelago provided a reachable destination halfway to his Asian target.[729] Thus, the first European discovered the New World by accident, using unrealistic and misleading maps!

Proof That Earth is Round

In 1519, Ferdinand Magellan and five ships departed from Seville, Spain, searching for a direct sea route to the spice islands of Indonesia.[730] Europeans prized these islands for their production of mace, nutmeg, cloves, and pepper, more valuable at the time than gold. Magellan's ships crossed the Atlantic Ocean and discovered a navigable sea route between mainland South America and Tierra del Fuego to the south, a passage known today as the Strait of Magellan. The expedition then crossed the Pacific and arrived at the Philippine islands—1,086 miles from Indonesia—where natives killed Magellan with poisoned arrows.

His second-in-command continued the expedition, circumnavigating the globe and limping back to Seville. Only one boat out of five and 18 of an original 260 sailors survived the voyage.[731]

---

[728] Charles C. Mann, *1493: Uncovering the New World Columbus Created* (New York: Vintage Books, 2012), 48.
[729] Mann, *1493*, 31.
[730] Laurence Bergreen, *Over the Edge of the World: Magellan's Terrifying Circumnavigation of the Globe* (Boston: Mariner Books, 2004).
[731] Bergreen, *Over the Edge of the World*, 403.

However, their circumnavigation is remembered as one of the great expeditions in seagoing history.

Earth, they proved beyond doubt, was truly round.

Darwin's Doubt

"Now faith is confidence in what we hope for and assurance about what we do not see…By faith we understand that the universe was formed at God's command, so that what is seen was not made out of what was visible." Hebrews 11:1–3 (NIV).

The English theologian and moral philosopher William Paley (1743–1805) lived by Paul's words in Hebrews.[732] Paley believed that the world was a blessed place, reflecting the goodness of nature and the goodness of the God who created it. He saw no difference between the human body and a fine Swiss watch, believing both reflected the skills of a great designer. His most successful work was *Natural Theology*, which presented in a clear and lucid manner all the evidence for the existence of God. He believed the universe is designed because it has complexity and purpose.[733]

Charles Darwin initially found Paley's arguments compelling, taking Paley's arguments to heart and writing: "I do not think I hardly ever admired a book more than Paley's *Natural Theology*: I could almost formerly have said it by heart."[734]

---

[732] William Paley, *Natural Theology* (Oxford: Oxford University Press, 2008), 236.
[733] Paley, *Natural Theology*, 280.
[734] Hanne Strager, *A Modest Genius: The Story of Darwin's Life and How His Ideas Changed Everything* (self-pub., CreateSpace, 2016), 3.

Then, at twenty-two years of age, Darwin was hired to be the naturalist on the HMS *Beagle*, remaining onboard from 1831 to 1836.[735] His travels on the *Beagle* provided Darwin with abundant specimens and opportunities for research, and his worldview began to change. He began to doubt that species were immutable,[736] focusing instead on changes in species induced by natural selection.[737] To Darwin, natural selection produced the benefits of adaptation, and at the same time removed the need for a designer.

Death of Annie

At the end of June 1850, Darwin's nine-year-old daughter Annie, a precious family member and a great comfort to her father, fell sick, suffered a painful illness, and died. The impact on Darwin's faith was profound.[738] After a twenty-year delay and four years after Annie's death, he published his seminal work, *On the Origin of Species by Means of Natural Selection, or the Preservation of Favoured Races in the Struggle for Life.*[739]

Darwin's faith began to falter. He would go for a walk on Sundays while his family attended church. Though reticent about his religious views, in 1879 he responded that he had never been an atheist in the sense of denying the existence of a god. But he declared a less-than-enthusiastic worldview: "An agnostic would be the most correct description of my state of mind."[740]

---

[735] Strager, *A Modest Genius*, 12.
[736] Strager, *A Modest Genius*, 82.
[737] Strager, *A Modest Genius*, 72.
[738] Strager, *A Modest Genius*, 85.
[739] Charles Darwin, *On the Origin of the Species* (London: John Murray, 1859).
[740] "To John Fordyce 7 May 1879," Darwin Correspondence Project, University of Cambridge, accessed XXX

Aftermath

Charles Kingsley, a Christian socialist country rector and novelist, gave *On the Origin of Species* a positive review, declaring "It is just as noble a conception of Deity, to believe that He created primal forms capable of self-development…as to believe that He required a fresh act of intervention to supply the lacunas which He Himself had made."[741] For the second edition, Darwin added these lines to the last chapter, with attribution to "a celebrated author and divine." But Darwin began to believe that nature, not divine planning, controlled the journey of life. He allowed a new visibility, namely natural selection, to work in the world, to alter his understanding of God.

The Impact of Natural Selection

Examples of natural selection have become well documented by biological science. In response to a change in food source, finches may develop longer beaks to capture insects, or shorter and tougher beaks to feed on seed.[742] Arctic rabbits change the color of their fur when snow covers their landscape.[743] Non-poisonous king snakes

---

[741] "From Charles Kingsley 18 November 1859," Darwin Correspondence Project, University of Cambridge, accessed XXX, https://www.darwinproject.ac.uk/letter/?docId=letters/DCP-LETT-5673.xml.

[742] Sangeet Lamichhaney et al., "Evolution of Darwin's Finches and Their Beaks Revealed by Genomic Sequencing," *Nature* 518 (February 2015): 371–375, https://doi.org/10.1038/nature14181.

[743] "Arctic Hare," National Geographic, accessed XXX, https://www.nationalgeographic.com/animals/mammals/facts/arctic-hare.

take on the appearance of poisonous coral snakes, while moths change their color in response to the color of a twig.[744]

Sacred Science Has Given Us Greater Vision

Today, do we see natural selection as an independent force in the world? Or do we place these adaptations in a more modern perspective? When a beak changes its morphology, what genes do we see at work? Researchers have identified the gene that changes the length and shape of a beak.[745] They have also identified the genes that induce morphological changes in fish in response to the presence of predators.[746] Evolutionary biologist David Reznick at the University of California has reported that some species of fish can make these morphologic changes in a mere six to eight generations.[747]

Darwin saw the forces of natural selection at work in the world, and it made him doubt. His doubt would become humanity's malaise for the next 150 years. At the middle of the twentieth century, our forefathers faced the fearsome power of nuclear war, accompanied by a dismal backdrop of atheistic Darwinism. They began to dismiss Paul's declaration of faith: "For since the creation of the world God's invisible qualities—his eternal power and divine nature—have been clearly seen, being understood from what has been made, so that people are without excuse." Romans 1:20 (NIV).

---

[744] Amy Eacock et al., "Adaptive Colour Change and Background Choice Behaviour in Peppered Moth Caterpillars is Mediated by Extraocular Photoreception," *Communications Biology* 2 (2019): article 286, https://doi.org/10.1038/s42003-019-0502-7.

[745] Lamichhaney et al., "Evolution of Darwin's Finches."

[746] Michael Traugott et al., "Fish as Predators and Prey: DNA-Based Assessment of Their Role in Food Webs," *Journal of Fish Biology* 98 no. 2 (May 2020): 367–382, https://doi.org/10.1111/jfb.14400.

[747] Jane Braxton Little, "Rapid Evolution Changes Species in Real Time," *Discover*, last updated April 22, 2020, https://www.discovermagazine.com/planet-earth/rapid-evolution-changes-species-in-real-time.

Darwin discovered a little science; he had made an important observation that, ironically, made him doubt. The antidote would be centuries in the making, and what form would it take? It would arrive as sacred science. Light creates quarks, while a Higgs mechanism gives them mass, and gluons shape them into the constituents of the atom. Great mathematicians retrieve just-right equations from a mathematical landscape on a higher dimension while supercomputers identify a library for life. Now, next-generation sequencing identifies the genes responsible for "natural selection."

Darwin should have taken Kingsley's words to heart: yes, it was "just as noble a conception of Deity, to believe that He created primal forms capable of self-development." God gave His creation free will. But He also gave His creatures a library of sequences and a nursery in their genome to accomplish divine intentions—creatures who would not only read and understand His equations, but who would also celebrate their divine source.

Darwin should have remembered these additional words from Paul: "Now faith is confidence in what we hope for and assurance about what we do not see."[748]

---

[748] Hebrews 11:1 (NIV)

Cracking the Code

Swiss chemist Friedrich Miescher first identified what he called "nuclein" (DNA) inside human white blood cells in 1869.[749] He had initially planned to isolate the protein components of white blood cells but discovered a nuclear substance with a much higher phosphorous content resistant to proteolysis (protein digestion). Miescher had stumbled upon DNA, but his peers dismissed his finding as an artifact produced by "contamination."

More than fifty years passed before scientists appreciated the significance of Miescher's discovery. A 1961 account of nineteenth-century science mentioned Charles Darwin thirty-one times but failed to mention Miescher even once. Of the four ingredients in a cell (proteins, lipids, polysaccharides, and nucleic acids), DNA was the only one that biologists could attribute to a single scientist at a single place and specific date. Yet, Miescher's discovery went unrewarded.[750]

---

749 Sophie Juliane Veigl, Oren Harman, and Ehud Lamm, "Friedrich Miescher's Discovery in the Historiography of Genetics: From Contamination to Confusion, from Nuclein to DNA," *Journal of the History of Biology* 53 (2020): 451–484, https://doi.org/10.1007/s10739-020-09608-3.
750 Veigl, Harman, and Lamm, "Friedrich Miescher's Discovery."

Grasping the significance of DNA would require much more research. In the 1950s, there was a dramatic race to discover the structure of DNA. Three teams were hot on the trail:[751]

1. Linus Pauling, of protein chemistry fame
2. Rosalind Franklin and Maurice Wilkins, who had perfected the technique of X-ray crystallography of molecules
3. James Watson and Francis Crick, young investigators at the University of Cambridge's Cavendish Laboratory

Early oddsmakers would have backed Pauling as the probable winner. His peers considered Pauling the founder of molecular biology following his discovery of the folded structure of proteins. His work with proteins went beyond a single discovery: he won a Nobel Prize for his entire body of work, including the nature of the chemical bond and the structure of the protein hemoglobin. Pauling even identified the cause of sickle cell anemia: a mutation makes sickle cell hemoglobin less soluble and more prone to deform into a "sickle" shape.[752]

Emergency Committee

But following the destruction of Hiroshima and Nagasaki, Pauling joined Einstein in campaigning against nuclear testing. Einstein, Pauling, and seven other scientists formed the Emergency Committee of Atomic Scientists.[753] Given the political environment, the US government did not appreciate Pauling's activism. The

---

[751] Matthew Cobb, *Life's Greatest Secret: The Race to Crack the Genetic Code* (New York: Basic Books, 2015).

[752] Thomas Hager, *Force of Nature: The Life of Linus Pauling* (New York: Simon and Schuster, 1995), 128.

[753] Hager, *Force of Nature*, 143.

FBI considered him a national security risk and blocked his travel outside the United States.

Had he completed a planned trip to London, Pauling, like Jim Watson, might have been among the first to see Rosalind Franklin's X-ray photo of DNA suggesting a helical structure.

At a conference in the spring of 1951 at the Zoological Station in Naples, Watson attended a presentation on the molecular structure of DNA and immediately appreciated the significance of Franklin's X-ray photographs. Notes from Franklin's lab books would indicate that she knew that DNA was a double helix. She also surmised that the arrangement of nucleotide bases along each strand was complementary, enabling the molecule to replicate. Even more, her notes also demonstrated that she had glimpsed the most incredible secret of DNA: the sequence of bases carried the genetic code.[754]

Based on his memory of Franklin's presentation, Watson and Crick put together a model of DNA and invited Wilkins and Franklin to see the model. Franklin quickly threw cold water on the project.

Watson's memory of her presentation was faulty, and their model was completely wrong.[755]

Watson and Crick needed precise observations from Franklin's X-ray crystallography to better understand the distances and relationships between the components of DNA. Franklin unwittingly included those numbers in a brief informal report to Max Perutz of Cambridge University. In February 1953, Perutz passed the information to Lawrence Bragg, the director of the Cavendish Laboratory, and Bragg passed it on to Watson and Crick.

---

[754] Cobb, *Life's Greatest Secret*, 124.
[755] James D. Watson, Andrew Berry, and Kevin Davies, *DNA: The Story of the Genetic Revolution* (New York: Knopf, 2017), 45.

Correcting the Model

Crick could now complete his calculations. In the middle of March 1953, Watson and Crick invited Wilkins and Franklin to review another model, and they immediately agreed the model must be correct.[756]

Together, they decided to publish the model as the work of Watson and Crick, while Wilkins and Franklin would publish the supporting data in separate papers. On April 25, there was a party to celebrate the publications. Franklin did not attend.[757]

James Watson, Francis Crick, and Maurice Wilkins received the Nobel Prize in Physiology or Medicine in 1962 for unraveling the structure of DNA.[758] Franklin had died four years earlier of ovarian cancer, possibly induced by X-ray exposure during her analysis of DNA.

Beyond DNA Structure to a Deeper Understanding of Nucleosides and Proteins

Living cells are the sole source of nucleotides, but nucleosides (a base attached to a sugar without the phosphate linkage) are widespread in nature.[759] Viruses often use nucleosides to reproduce their code. Pharmaceutical companies have created many nucleosides to act like Trojan horses, inserting themselves into the viral sequence

---

[756] Cobb, *Life's Greatest Secret*, 125.
[757] Cobb, *Life's Greatest Secret*, 126.
[758] Watson, Berry, and Davies, *DNA*, 57.
[759] Peter Yakovchuk, Ekaterina Protozanova, and Maxim D. Frank-Kamenetskii, "Base-Stacking and Base-Pairing Contributions into Thermal Stability of the DNA Double Helix," *Nucleic Acids Research* 34, no. 2 (January 2006), 564–574, https://doi.org/10.1093/nar/gkj454.

to block its replication. That strategy has cured hepatitis C and provides antiviral treatment for HIV.[760]

Manufacturing Nucleosides

Scientism presumes that DNA somehow self-assembled on early Earth. Orson Wedgwood, PhD, a research scientist involved in drug development for Alzheimer's and other diseases, begs to disagree. In a book titled *DNA: the Elephant in the Lab*, he outlines the challenges of controlling a synthetic process, even in the disciplined environment of a research lab:[761]

1. Oxygen can destroy the chemical reactions: it was necessary to flush the glass flask multiple times with pure nitrogen to remove all oxygen.

2. Curiously, water was not always the best solvent. Sometimes it was necessary to use a variety of different solvents, often avoiding water.

3. The process depended on beginning with enough pure reagents at suitable concentrations to avoid cross-reactions.

4. Scientists had to control temperature within a tight range.

5. The mixture had to be stirred nonstop to ensure even distribution of the starting chemicals.

6. Some reactions needed an acid environment. Others only proceeded if the acid-base balance was alkaline.

7. Reactions often required catalysts to speed up chemical reactions.

---

[760] *LiverTox: Clinical and Research Information on Drug-Induced Liver Injury* (internet), National Institute of Diabetes and Digestive and Kidney Diseases, last modified August 8, 2014, https://www.ncbi.nlm.nih.gov/books/NBK547852/.

[761] Orson Wedgwood, *DNA: The Elephant in the Lab—The Truth about the Origin of Life* (self-pub., Wedgwood Publishing, 2019), 57.

8. Scientists had to stop reactions at precisely the right time and purify the products thoroughly to remove unwanted contaminating products.

It is challenging to imagine where or how this complex process might have happened in the uncontrolled conditions of early Earth.

Exploring the Hyperspace of Protein Folds

Scientists have cracked the code for DNA but face an even greater challenge.

Life takes advantage of more than two hundred million amino acid sequences,[762] but protein scientists have only determined the structure of 1 percent of that total. Bench top machines can readily determine the sequence of amino acids, identifying which amino acid out of twenty types forms each bead along the length of a protein. But understanding a protein fold is a different matter entirely. Scientists have been able to see a small number of protein folds using X-ray crystallography,[763] but this process is time-consuming and imprecise.

An innovative approach has delivered a breakthrough. Science is using powerful computers and artificial intelligence to predict the shape of a protein's fold. Using data from thoroughly studied protein folds, scientists are using machine learning to predict the folding of other proteins.

Protein folding reflects a balance of various atom-scale forces, such as the interaction among chemical groups on adjacent amino

---

[762] Alexander Rives et al., "Biological Structure and Function Emerge from Scaling Unsupervised Learning to 250 Million Protein Sequences," *PNAS* 118, no. 15 (April 2021): article e2016239118, https://doi.org/10.1073/pnas.2016239118.

[763] Gregory S. Girolami, *X-Ray Crystallography* (Melville, NY: Universe Science Books, 2016).

acids and the interaction of an amino acid with its chemical environment. Does an amino acid love water, for instance? If so, it is hydrophilic. Or is it hydrophobic, avoiding water at any cost? That single feature can have a significant impact on the twists and folds of a protein.

What would science gain by accessing information about protein folding? The ability to decipher a protein's fold could rapidly expand our understanding of the "nanomachines" animating the activities of the cell. It could revolutionize our approach to diseases such as Alzheimer's and cancer. It could identify the perfect target for a viral vaccine, allowing us to quickly escape the next pandemic.

A laboratory has worked with a team of more than one hundred machine learning computers to develop the algorithms required to analyze protein folding. Scientists have trained those computers, with information about the 170,000 or so known protein folds. With that knowledge at hand, this artificial intelligence (AI) process has identified new protein folds with a success rate of more than 90 percent.[764]

For the first time in the history of science, human beings have an entirely new and fast way to analyze a protein's folding pattern and identify its "active site." Hopefully, this ability will soon translate into healthcare breakthroughs.

Significance of AI

As AI advances, one might read the chapters of this book and wonder: are we the product of an AI platform using unimaginably

---

[764] Robert F. Service, "The Game Has Changed. AI Triumphs at Protein Folding," *Science* 370, no. 6521 (December 2020): 1144–1145, https://doi.org/10.1126/science.370.6521.1144.

advanced supercomputers? We should respond with a straightforward question. Would AI, however advanced:

1. identify a chosen people;
2. inspire messianic dreams;
3. declare love the priority of creation;
4. take human form and challenge the misplaced priorities of
   temple worship;
5. defy the world's most powerful empire; or
6. willingly submit to death on a cross?

Would AI be able to glorify that sacrifice by demonstrating its
power over death?

As AI advances, the vital importance of our religious beliefs
and scripture will come into clearer focus.

Regulatory Genes on Continents Apart

Two hundred million years ago, the Earth possessed a single super continent called Pangea. Slowly over time, tectonic plates separated Pangea into Australia and North America and provided separate platforms for evolution.[765] Would similar creatures evolve in worlds separated by vast oceans and deep time?

Indeed, they have, and different modes of newborn development neatly divide the results.

Marsupials

The tiny, undeveloped young Australian marsupials crawl into an external pouch on the underside of their mother's body and complete their development peeking out from the top of the pouch.

Placentals

Placental species in North America complete their development inside their mother's womb.

---

[765] Martin Ince, *Continental Drift: The Evolution of Our World from the Origins of Life to the Far Future* (New York: Blueprint Editions, 2018).

Despite having no common ancestor, numerous animals from both continents share near-identical features. The Northern flying squirrel and the Australian super glider have small furry bodies, large eyes, and characteristic flaps of skin that extend between the front and back legs. You would be hard-pressed to tell the difference.[766]

The North American mole and a marsupial mole have near-identical features, both adapted for a life underground. Their features include enormous claws for digging and eyes that are covered by fur.

The list of near-identical life-forms, among others, includes rabbits, anteaters, and mice.

Australian Wolf

The Tasmanian tiger or "wolf" (technically called a "thylacine," or wolf with a pouch) was once the top canine on the Australian continent.[767] Unfortunately, the wild dog (the dingo) introduced by European immigrants was a more efficient hunter. Out competed, and suffering from a shrinking gene pool, the last Tasmanian wolf died in a zoo in 1936.[768]

The Tasmanian version of the wolf carried its babies in a pouch like a kangaroo. Aside from this difference, it shared strikingly similar anatomic features with its North American counterpart.

---

[766] Cornelius Hunter, "The Real Problem with Convergence," Evolution News and Science Today, May 25, 2017, https://evolutionnews.org/2017/05/the-real-problem-with-convergence/.

[767] Charles Y. Feigin, Axel H. Newton, and Andrew J. Pask, "Widespread *cis*-Regulatory Convergence between the Extinct Tasmanian Tiger and Gray Wolf," *Genome Research* 29 (2019): 1648–1658, https://doi.org/10.1101/gr.244251.118.

[768] Asher Elbein, "Tasmanian Tigers are Extinct. Why do People Keep Seeing Them?" *New York Times*, March 10, 2021, https://www.nytimes.com/2021/03/10/science/thylacines-tasmanian-tigers-sightings.html.

Despite having no common ancestor in over one hundred million years, as mentioned briefly in the introduction to chapter 13, both species had nearly identical paws and skulls. In fact, professors of anatomy at Oxford University often challenge their students to distinguish a dog's skull from that of a Tasmanian wolf during practical exams. Most students fail the challenge.[769]

The shape of the Tasmanian tiger's skull was virtually identical to the skull of the North American red fox and gray wolf, but completely different from the skull of its closest marsupial relative, an anteater called the numbat. (Who chose such a name?)

Regulatory Genes Determine the Timing of Gene Expression

For decades, scientists have assumed that the similarity between the Tasmanian tiger and the North American wolf reflected the power of natural selection. These animals faced similar challenges, the story goes, to find food and stay alive. Natural selection, Darwinists propose, worked its magic and shaped the skull of these "wolves."[770]

Sequencing Surprise

Welcome to a classic example of modern DNA sequencing informing the science of life. Scientists have succeeded in sequencing the genome of a Tasmanian pup preserved in alcohol as a museum specimen. What did they find? Near-identical regulatory genes

---

[769] Richard Dawkins, *The Ancestors Tale: A Pilgrimage to the Dawn of Evolution* (Boston: Mariner Books, 2004), 277.
[770] Marc Tollis, "Case Studies of Convergent Evolution: of Wolves and Thylacines," Anolis Tollis (blog), March 5, 2014, https://anolistollis.wordpress.com/2014/03/05/case-studies-of-convergent-evolution-of-wolves-and-thylacines/.

shaped the skull of both the thylacine and American wolves.[771] Regulatory codes channeled two unrelated species along the same developmental path, giving them identical jaws and teeth.

How did the same regulatory sequences find their way into the genome of these distantly related creatures? Natural selection may have played a role in the process, promoting the transcription of one set of genes over another, but that begs a deeper question. What nominated these genes in the first place, assembling them in the genome of distantly related species millions of years in the past?

On the discredited Darwinian tree of life, the thylacine should share a limb with the numbat.

Regulatory genes have chopped down Darwin's tree.

---

[771] Charles Y. Feigin et al., "Genome of the Tasmanian Tiger Provides Insights into the Evolution and Demography of an Extinct Marsupial Carnivore," *Nature Ecology and Evolution* 2 (2018): 182–192, https://doi.org/10.1038/s41559-017-0417-y.

Mathematical and Man-Made Libraries

The codes for life are as mathematical as the Higgs equation. Yes, DNA sequences become physical as nucleotides attach to the backbone of DNA. Yes, the sequences of amino acids become physical as proteins collapse into three-dimensional shapes. But before they take on material structure, DNA and protein sequences populate a mutational landscape. They live in the same immaterial domain as the Higgs equation.

DNA is a tightly woven, highly efficient language that follows extremely specific rules.[772] Researchers believe its alphabet, grammar, and overall structure reflect the underlying presence of a beautiful set of mathematical functions.[773]

Both DNA, RNA, and proteins can be expressed in letter codes. For example, the initial three mRNA codes for amino acids

---

[772] Jean-Claude Perez, "Codon Populations in Single-Stranded Whole Human Genome DNA are Fractal and Fine-Tuned by the Golden Ratio 1.618," *Interdisciplinary Sciences: Computational Life Sciences* 2 (2010): 228–240, https://doi.org/10.1007/s12539-010-0022-0.

[773] David Swigon, "The Mathematics of DNA Structure, Mechanics, and Dynamics," in *Mathematics of DNA Structure, Function and Interactions*, The IMA Volumes in Mathematics and its Applications vol. 150, eds. Craig John Benham et al. (New York: Springer, 2009), 293–320, https://doi.org/10.1007/978-1-4419-0670-0_14.

in chicken ovalbumin can be expresses as: CCU, UUA, GCA (mRNA)[774]; and as Proline, Leucine, Alanine (amino acids); and amino acids as three letters: Pro, Leu, Ala; and, finally, amino acids as a string of single letters, where each letter represents an amino acid: P, L, A.

Researchers have also constructed mathematical models of human metabolisms,[775] and other scientists are using a mathematical application of information theory to identify and predict mutations in SARS-CoV-2, commonly known as COVID-19.[776]

It is hardly a stretch to believe that our superintelligent divine Creator could organize the sequences of life, like the mathematics of the universe, in a library of "ideal forms."

Legendary Libraries

Created in the image of God, human beings have repeatedly attempted to represent their world through symbols, organizing their thoughts as etchings on clay tablets, hieroglyphs on papyrus and later, as letters on tablets and paper. Widely separated cultures have exerted great effort to collect all the known writings of their era in libraries of world renown.

The Royal Library of Ashurbanipal, named after the last great king of the Assyrian Empire, contained a collection of more than thirty thousand clay tablets and fragments from the seventh century

---

[774] McReynolds, L., O'Malley, W., Nisbett, A. D. et. al. (1978). Sequence of chicken ovalbumin mRNA. Nature, 273:723.

[775] T. Pearson et al., "A Mathematical Model of the Human Metabolic System and Metabolic Flexibility," *Bulletin of Mathematical Biology* 76 (2014), 2091–2121, https://doi.org/10.1007/s11538-014-0001-4.

[776] Melvin M. Vopson and Samuel C. Robson, "A New Method to Study Genome Mutations Using the Information Entropy," *Physica A: Statistical Mechanics and its Applications* 584 (2021): article 126383, https://doi.org/10.1016/j.physa.2021.126383.

BC. It was located in the ancient Mesopotamian city of Nineveh. In 612 BC, an invading horde, including Babylonians from what is now modern-day Iraq, Scythians from what is now known as southern Siberia, and the Medes from Iran, destroyed Nineveh and the library. In 1849, Layard recovered tablets that survived the invaders' fires, now on display in the British Museum.[777]

The Great Library of Alexandria, Egypt held as many as seven hundred thousand documents from Greece, Persia, Egypt, India, and other regions, making Alexandria the capital of knowledge and learning in 300 BC. Alexandrian scholars focused on rhetoric, law, epics, tragedy, comedy, lyric poetry, history, medicine, mathematics, and natural science.[778]

Seventy scholars completed the Greek translation of the Old Testament, the "Septuagint," in the library of Alexandria in the mid-third century BC.

Much later, between the eighth and thirteenth centuries, Baghdad emerged as the capital of the Abbasid Caliphate. Scholars flowed into Baghdad and developed a vibrant academic community. Their library, referred to as the "Storehouse of Wisdom," included works devoted to philosophy, mathematics, medicine, astronomy, and optics. But in 1258, Mongols sacked Baghdad and destroyed the library, burning the world's most extensive collection of precious documents.[779]

The Chinese have also attempted to document their culture and knowledge in written records, sometimes with surprising backlash.

---

[777] Grant Frame and A. R. George, "The Royal Libraries of Nineveh: New Evidence for King Ashurbanipal's Tablet Collecting," *Iraq* 67, no. 1 (2005): 265–284, https://doi.org/10.1017/S0021088900001388.

[778] Justin Pollard and Howard Reid, *The Rise and Fall of Alexandria: Birthplace of the Modern World* (New York: Viking Adult, 2006).

[779] Jim Al-Khalili, *The House of Wisdom: How Arabic Science Saved Ancient Knowledge and Gave Us the Renaissance* (New York: Penguin, 2012), 226.

The Chinese phrase translated to the "burning of books and bury-
ing of scholars" refers to an unfortunate event in 213 B.C. when
emperor Quin Shi Huang ordered the burning of texts and burial of
460 Confucian scholars alive.[780] Leaders of present-day China have
funded notable libraries including The National Library of China,
the Shanghai Municipal Library, and Peking University Library.

Today, the Library of Congress is the oldest federal cultural
institution in the United States and one of the largest libraries in
the world, containing almost 126 million books and manuscripts,
among millions of other items. But in August 1814, British troops
burned the Library of Congress, including its collection of three
thousand volumes.[781] To restore its collection, the library purchased
Thomas Jefferson's personal collection of 6,487 books. Unsurpris-
ingly, another fire in 1851 destroyed many of Jefferson's books.[782]

Human libraries, it seems, are always at risk of destruction.
Thankfully, the mathematical libraries storing equations for the
universe and genetic codes for life are immune to human assault.
Their stacks are always open, inviting minds and mutations to
explore their priceless treasures.

---

[780] Lois Mai Chan, "The Burning of the Books in China, 213 BC," The Journal of
Library History 7, no. 2 (April 1972): 101–108, https://www.jstor.org/
stable/25540352.

[781] John Y. Cole, *America's Greatest Library: An Illustrated History of the Library of
Congress* (Lewes, UK: GILES, 2018), 137.

[782] Cole, *America's Greatest Library*.

Energy and Elements for Life: Additional Evidence for Design

Remarkably, the same hydrogen proton that provides the deuterium and tritium for nuclear fusion in the stars generates the energy for life.[783]

Resembling tiny cigars or footballs, mitochondria float by the hundreds in every cell.[784] Mitochondria have an outer membrane like the nuclear membrane surrounding our DNA. But mitochondria also have an inner membrane with sub-compartments called "cristae," whose multiple folds maximize the membrane surface. The outer membranes and the cristae create two separate liquid compartments, like a hot tub in the middle of a swimming pool:

1. There is an outer compartment between the mitochondrial membrane and the cristae (the swimming pool).

---

[783] Eric Smith and Harold J. Morowitz, *The Origin and Nature of Life on Earth: The Emergence of the Fourth Geosphere* (Cambridge: Cambridge University Press, 2016), 327.

[784] Lynn Margulis, *Symbiotic Planet. A New Look at Evolution* (New York: Basic Books, 1999).

2. There is an innermost compartment—the metaphorical hot tub—called the "mitochondrial matrix," an innermost soup enclosed by the convolutions of the cristae.

Reactions in the matrix produce the chemical energy for life, including the energy required for protein synthesis, muscle movement, or even thoughts in our brain.

Wondrous Proteins

Unique proteins live on these cristae and perform amazing feats of biochemical engineering. The following is a simplified description of the process.

During the chemical breakdown of food, our cells create electron-rich molecules containing a hydrogen atom (NADH or Vitamin B3 are good examples).[785] Nicotinamide Adenine Dinucleotide plus hydrogen (NADH) helps all the cells in the body produce energy by delivering electrons to the electron protein chain and thus to the interior of the mitochondrion, the batteries that live inside our cells.

Electron-transport proteins straddle the membranes of the mitochondrial cristae and steal every available electron.[786] Oxygen in the inner matrix then absorbs those electrons like a sponge, producing water and carbon dioxide.[787] This creates a concentrated pool of protons in the outer pool as electron-transport proteins steal electrons and leave protons on their own.

---

[785] George D. Birkmayer, *NADH: The Biological Hydrogen* (Nashville, TN: Basic Health Publications, 2009).

[786] Smith and Morowitz, *The Origin and Nature of Life*, 278.

[787] Philip Nelson, *Biological Physics: Energy, Information, Life* (Philadelphia: Chiliagon Science, 2003), 463.

Now, one of the most spectacular proteins in our body makes use of the proton pool. A protein called "ATPase" inserts a turbine-like rotor in the membrane between the pool and the hot tub.[788] ATPase invites free protons to flow through its protein core, and the flow of protons turns the ATPase rotor 130 revolutions per minute.

Creating Chemical Energy with Each Rotation

Like an automobile crankshaft injecting fuel into the cylinder of an engine, the ATPase rotor injects an extra phosphate bond into a chemical called "adenosine diphosphate" (ADP). ADP, with two phosphate bonds, becomes adenosine triphosphate (ATP), with three. The energy in that new phosphate bond provides the energy for the activities of life including the movement of muscles and the flow of thoughts in our brain.

Energy Takes Many Forms

Energy (the ability to do work) takes many forms.[789] Energy can be mechanical, thermal, nuclear, electromagnetic, and solar. But energy can also be chemical, and ATPase stores energy in chemical bonds. We can understand the chemical energy of ATP with a simple metaphor. ADP is a Fourth of July sparkler that will barely sparkle. But ATP is a brightly burning sparkler that can act like a magic wand, using the energy in its third phosphate bond to underwrite the activities of the cell.

---

788   Nelson, *Biological Physics*, 423.
789   "What is Energy?" US Energy Information Administration, last updated August 16, 2023, https://www.eia.gov/energyexplained/what-is-energy/forms-of-energy. php.

Another metaphor may make the workings of the mitochondrion more understandable. Imagine high school seniors waiting for tickets to the prom. Unbeknownst to the boys (the protons), an usher has slipped free passes to the girls (the electrons), who all head into the dance. Suddenly realizing they are part of a male-only crowd, the left-behind "protons" rush to enter the dance hall, rapidly turning the ATP turnstile. The rotating turnstile pays a precious dividend: every rotation produces several new molecules of ATP.

Vast quantities of proteins separate electrons from protons, creating the proton gradient that turns the rotor of ATPase. The mitochondrial membranes in each human body would cover four football fields. More than 10^{21} protons flow across our mitochondrial membranes every second.[790] *We generate half our body weight in fresh ATP every day.*

For millennia, humans have used rotating water wheels to produce energy for the irrigation of crops and the grinding of grains. God's mitochondria beat us to the power of the rotor by hundreds of millions of years.

Other Elements for Life

Cosmic printing filled the heavens with galaxies and stars, but God's printing plans were hardly complete. Each star represents a printing platform of its own, using nuclear fusion to convert simple hydrogen, a single proton, into heavier elements designed to meet the needs of life.

---

[790] James P. Bennett and Isaac G. Onyango, "Energy, Entropy and Quantum Tunneling of Protons and Electrons in Brain Mitochondria: Relation to Mitochondrial Impairment in Aging-Related Human Brain Diseases and Therapeutic Measures," *Biomedicines* 9, no. 2 (February 2021): 225, https://doi.org/10.3390/biomedicines9020225.

Oxygen

Most of us assume oxygen has magical powers to conduct respiration on its own. As described above, the truth is more complex. Oxygen sponges up the electrons flowing into the inner mitochondrial compartment, removing electrons from the field of battle.[791] Think of oxygen as a bouncer in front of an old-time western saloon. As the drunks pour out of the door, oxygen carries them away so that new customers find room in the bar.

What would happen without oxygen? Electrons accumulate outside the cristae, stopping electron transport. Protons would reclaim their electrons and lose interest in the ATPase turnstile. The production of ATP would stop. How long can you go without oxygen? Deprived of oxygen, you lose consciousness within four minutes.[792]

We Should Give Our Oxygen-Containing Atmosphere Independent and Due Consideration

The atmosphere of Earth meets a variety of needs:[793]

1. It provides the pressure required for liquid water to exist on the planet's surface.
2. It protects us from ultraviolet solar radiation.
3. It retains heat and warms the surface through a greenhouse effect, minimizing temperature swings between night and day.

---

[791] Nelson, *Biological Physics*, 459.
[792] Bethany Cadman, "Everything You Need To Know About Anoxia," Medical News Today, January 11, 2018, https://www.medicalnewstoday.com/articles/320585.
[793] Paul I. Palmer, *The Atmosphere: A Very Short Introduction* (Oxford: Oxford University Press, 2017).

4. It provides the oxygen required for cellular respiration, as described above.

Oxygen (O2) is a unique molecule, the second-most abundant element in the sky and the most abundant element in the Earth's crust.

Carbon: The Chain and Ringmaster

Carbon (C) is the "backbone" of life. It bonds with itself and a wide variety of other elements to form essential molecules, such as DNA and chains of amino acids called "proteins."

Electrons orbit the nucleus of the atom in shells. The innermost shell is limited to two electrons. The next shell has the capacity for eight. Carbon has six protons and thus six electrons. Two of those electrons orbit in the innermost shell. The remaining four electrons utilize half of the eight slots available in the outer shell.

Four Free Electron Slots Facilitate Carbon's Deals with Its Neighbors

This distribution of electrons has a profound effect on the chemistry of life. Carbon has four available empty slots in that second shell to make a variety of molecules essential for life. One obvious option: share four electrons with another atom of carbon. Carbon can make long chains of carbon as well as carbon rings.[794]

During our discussion of quantum mechanics, we reviewed Fred Hoyle's insightful prediction that carbon would take an extraordinary quantum leap in the stars to turbocharge the fusion

---

[794] Robert M. Hazen, *Symphony in C: Carbon and the Evolution of (Almost) Everything* (New York: W. W. Norton and Company, 2019).

of carbon. As two helium atoms entangle to form beryllium, a third helium buzzes into the process, creating carbon-12 instead of beryllium-8. Knowing the critical role carbon plays in the structure and function of life, Hoyle predicted carbon's quantum energy levels in the stars before a lab confirmed his guess. A perfect quantum leap makes carbon one of the most common elements in the universe.[795]

Carbon atoms can form multiple stable bonds with other small atoms, including hydrogen, oxygen, and nitrogen. That opens the door to organic chemistry as carbon forms molecules in four major groups:

1. Carbohydrates (sugar and starch)
2. Lipids
3. Proteins
4. Nucleic acids (DNA and RNA)

Carbon constitutes 18 percent of our body.[796] Little wonder that God gave carbon the perfect quantum leap.

Carbon is the core component of organic chemistry, the chemistry of life, capable of forming millions of complex organic compounds.[797] It can bond firmly with itself to create chains of almost unlimited length. Because it can host four electron-sharing bonds, it naturally forms ring structures, creating molecules with well-defined three-dimensional shapes. The double helix of DNA and the folding of proteins depend on the shape-forming power of carbon. Carbon is essential to DNA, RNA, amino acids, and proteins.

[795] Evgeny Epelbaum et al., "Structure and Rotations of the Hoyle State," *Physical Review Letters* 109, no. 25 (December 2012): article 252501, https://doi.org/10.1103/PhysRevLett.109.252501.
[796] Michael Schirber, "The Chemistry of Life: The Human Body," Live Science, April 16, 2009, https://www.livescience.com/3505-chemistry-life-human-body.html.
[797] Hazen, *Symphony in C.*

Water: The Liquid of Life

We have visited the contributions of hydrogen and oxygen on their own. What happens when these atoms form a molecular combination called water (H2O)?

Oxygen and hydrogen create molecules of H2O. Water might appear to be the epitome of a simple molecule, just two hydrogen atoms attached to a single atom of oxygen. But the behavior and benefits of water reflect ingenious design. Two hydrogen atoms bond very strongly with oxygen. But rather than attaching in a straight line, they form an angle, giving the oxygen "side" of the water molecule a slightly negative electric charge. In contrast, the hydrogen "side" of the water molecule has a weakly positive charge.[798] Without this asymmetry, life would never have appeared on our planet.

Preventing Snowball Earth

Water is one of the very few liquids in the universe that floats in its frozen form. When water freezes, the positive sides of the molecules cozy up to the negative sides of adjacent molecules, creating some empty space. The result: solid water (ice) is less dense than liquid water—making it possible for ice to float. Ice floating on rivers and lakes prevents the freezing of water below.[799] If ice sank, every pond, lake, and river would be permanently frozen, denying life a foothold on the planet.

---

[798] Robert M. Hazen, *The Story of Earth: The First 4.5 Billion Years, from Stardust to Living Planet* (New York: Viking, 2012), 81.

[799] Guillermo Gonzalez and Jay W. Richards, *The Privileged Planet: How Our Place in the Cosmos Is Designed for Discovery* (Washington DC: Regnery Publishing, 2004), 33.

Water is a Precious Commodity

1. It is a "universal solvent."
2. It transports nutrients throughout the body.
3. It flows in rivers and waterfalls and releases vital elements from rocks.
4. It regulates body temperature, allowing humans to perspire and dogs to pant.
5. It lubricates Earth's tectonic plates.
6. It is essential for the irrigation of crops.
7. Water is also a symbol of purity, used for ritual washing in Christianity, Judaism, Hinduism, Islam, Shinto, Taoism, and other religions.

Water, not gold, will be the most valuable resource on an increasingly crowded Earth.[800]

---

[800] Giulio Boccaletti, *Water: A Biography* (New York: Pantheon Books, 2021).

Mutational Activity in "Neutral" (Non-Coding) DNA

A library of sequences for life in a higher dimension,[801] combined with the gathering of useful mutations in the quiet recesses of DNA in a process referred to as "preassembly," represent a new foundation for the origin of species compatible with divine intentions.[802]

Have scientists found evidence for the preassembly of new genes in non-coding DNA? In a vastly significant discovery, researchers have shown that "non-coding sequences can evolve into completely novel proteins."[803] These scientists may be witnessing preassembly in action.

[801] Andreas Wagner, *Arrival of the Fittest: Solving Evolution's Greatest Puzzle* (New York: Current, 2014).

[802] Fredric M. Menger, "An Alternative Molecular View of Evolution: How DNA was Altered over Geological Time," *Molecules* 25, no. 21 (November 2020): article 5081, https://doi.org/10.3390/molecules25215081.

[803] Matt Wood, "Genes That Evolve from Scratch Expand Protein Diversity," University of Chicago Medicine, March 10, 2019, https://www. uchicagomedicine.org/forefront/biological-sciences-articles/2019/march/ genes-that-evolve-from-scratch-expand-protein-diversity.

Mutations Perfectly Suited for Library Visits

For decades, evolutionary biologists have believed that there were only two ways a new gene might evolve. First, duplication might make an extra gene copy. The duplicate copy would be free to mutate and search for a new function. Second, when parental genomes unite to create the next generation through the merger of their DNA in a process called "recombination," pieces of genetic material might reshuffle and create new combinations and new genes. However, these two methods can only account for a small number of new proteins.

Scientists have confirmed a third way for new genes to appear—through mutational activity in the protective realm of non-coding DNA. As we have already noted, non-coding sequences represent 98 percent of the genome. Mutations in these non-coding sections of the genome would have no immediate impact on the host. But multiple mutations making repeat visits to the mutational library could gather vital genes for the future. One gene for a new function or body structure drifts into the genome, then another, and yet another, until the entire repertoire is available for production. Suddenly, the trilobite has eyesight and can visualize its prey.

Why Has "Junk" Been Preserved for Eons?

There is compelling evidence that this "junk" DNA plays a crucial role in life's journey. Studies have shown that generous portions of this so-called junk have survived in the genome for hundreds of millions of years.[804] This preservation over deep time suggests that

---

[804]   Emily S. Wong et al., "Deep Conservation of the Enhancer Regulatory Code in Animals," *Science* 370, no. 6517 (November 2020): article eaax8137, https://doi.org/10.1126/science.aax8137.

life views this junk as a treasure. These quiet DNA reserves may also be nothing less than the nursery for the "origin of the species."

Atheistic evolutionists have tried to show God the door. A library for life and a process called preassembly have confirmed what we have long suspected: God had a plan for the journey of life before the creation process began. He allows life to explore, but He had preconceived and preferred destinations. His library and the process of preassembly executed His commands to perfection, filling the oceans, the skies, and the land with a glorious array of creatures.

Preassembly of New Body Plans in "Junk" DNA

Imagine a dress shop that takes an unusual approach to its product line. Its proprietors make almost imperceptible changes to their dress from year to year. One year, they might change the pull tab on a zipper. Several years later, they modify a button. No matter how consumer tastes change, this clothing company stays true to minor variations, hoping customers will select their product.

Another dress shop does things differently. Savvy to the fickle nature of consumer taste, they invest in a designer warehouse far removed from the public's eye. A design team works in their warehouse to develop new dress plans from bodice to skirt. When the time seems right for a new design, they ship their patterns to production, and an entirely new dress takes a place on the racks.

If every dress shop were to work like the former, clients everywhere would fill their closets with look-alike dresses. Should dress shops choose the second model of operations, closets would contain bold new designs.

But dress shop number two must commit to a cloistered workshop and an operation that ignores specific timelines. Even more importantly, the owners of dress shop number two must dispatch their designers to fashion meccas several times a year to stay abreast of the best ideas.

Like the workings of a premium dress shop, neutral, non-coding DNA could provide the research and development laboratory for life. Relieved of the pressure for repetitive dress production, neutral genes could take multiple trips to the design library and shop for innovative ideas. When mutations have gathered for a winning new idea, the latest and greatest design makes a grand appearance.

Once the genes for eyes and claws have preassembled in the genome of Cambrian creatures, trilobites pursue their prey. Once the genes for sonar have preassembled in the genomes of bats and whales, sea creatures hunt in the dark ocean waters. Once the genes for the most complex brain in the universe have preassembled in the hominid genome, *Homo sapiens* draw images of animals on the walls of caves and place religious artifacts in the graves of their dead.

Mutational trips to the sequence library, followed by preassembly and supplemented by natural selection, have steered the journey of life, accomplishing divine intentions.

Conflicting Animal and Navy Sonar

Sound travels much faster through water than it does through air. A horizontal layer of water in the ocean acts as a waveguide for sound, allowing sound waves to travel thousands of miles.[805] Whales take advantage of this fact, communicating with one another over distances exceeding hundreds of miles.

Before ship noise polluted the oceans, fin whales emitted sounds that traveled up to thirteen thousand miles—a greater distance than the diameter of the Earth.[806]

Voracious Sperm Whales

Sperm whales eat about one ton of fish and squid per day. How do they hunt in the utter darkness of the deep ocean? At least a quarter of a sperm whale's length is dedicated to its nose, and that nose contains an elaborate sound-generating device. Sperm whales

---

[805] "What Is SOFAR?" National Oceanic and Atmospheric Administration, last updated June 16, 2024, https://oceanservice.noaa.gov/facts/sofar.html.

[806] Sophy Grimshaw, "Calls from the Deep: Do We Need to Save the Whales All Over Again?" *The Guardian*, December 31, 2020, https://www.theguardian.com/environment/2020/dec/31/calls-from-the-deep-do-we-need-to-save-the-whales-all-over-again.

produce a beam of sound from the front of the head like a search-light illuminating their prey.[807]

Sperm whales, the kings of the ocean, use echolocation to identify every meal. Scuba divers encountering a pod of sperm whales report the sensation of sound waves pulsating through their bodies. Fortunately, the whales can distinguish the difference in echoes produced by human divers and delicious giant squid.

Navy Sonar

Like the deep-diving sperm whale, the US Navy deploys active sonar, analyzing echoes to detect mines and other potential threats.[808] Naval investigators have long known that loud, low, and pure tones, often a single-frequency wave, free of distortion that course through ocean waters. Biologists have demonstrated that whales produce those sounds. They have also built a clear case that sonar threatens the health and livelihood of those whales.

Evidence of Stress

Beaked whales (rarely seen because they live in deep ocean waters) and blue whales rapidly move away from man-made sonar. Sonar sounds can disrupt feeding and even precipitate mass beaching.[809]

---

[807] Frants H. Jensen et al., "Narrow Acoustic Field of View Drives Frequency Scaling in Toothed Whale Biosonar," *Current Biology* 28, no. 23 (December 2018): 3878–3885.E3, https://doi.org/10.1016/j.cub.2018.10.037.

[808] Fred T. Erskine III, *A History of the Acoustics Division of the Naval Research Laboratory: The First Eight Decades 1923–2008*, Naval Research Laboratory, August 2013, https://apps.dtic.mil/sti/tr/pdf/ADA586269.pdf.

[809] Y. Bernaldo de Quirós et al., "Advances in Research on the Impacts of Anti-Submarine Sonar on Beaked Whales," *Proceedings of the Royal Society B* 286, no. 1895 (January 2019): article 20182533, https://doi.org/10.1098/rspb.2018.2533.

We are involved in a new cold war—between navy sonar and the echolocating leviathans of the seas.

The Wind Industry is Accelerating the War

The death of whales along the East Coast of the US has reached alarming proportions. Researchers have identified a correlation between increasing whale death and wind-industry boat traffic, including illegally loud sonar used to map the ocean floor. In their effort to identify the best location for ocean-based windmills, boat traffic and sonar are pushing whales into harm's way. A disturbing case in point: the dead body of a whale named "Faith" was identified floating at the surface at a wind turbine base.[810]

Toothed whales (dolphins, porpoises, and all other whales possessing teeth including sperm whales) have used echolocation for more than thirty million years. For many marine mammals, including beaked whales, blue whales, and humpback whales, seeing with sound is their handle on the world. Humankind has used sonar for just a few hundred years. We should minimize our intrusion into their deep ocean realm.

---

[810] Vivek Saxena, "Critics Vow Gov't Offshore Wind Scandal About to 'Explode'; State's Entire Fisherman's Board Bails on Biden-Backed Travesty, " Business and Politics, September 7, 2023, https://www.bizpacreview.com/2023/09/07/critics-vow-to-expose-corrupted-offshore-wind-scandal-states-entire-fishermans-board-bails-on-biden-backed-travesty-1393930/.

Continuing the Alien Search

The professional photographer Craig Foster spent a year of his life snorkeling in the kelp forests of the South African ocean.[811] His daily dives soon focused on an unexpected mission: developing a relationship with an octopus.

During his repeated visits to the world of the female mollusk, documented in the Netflix film *My Octopus Teacher*, he observed her using tools by gathering shells to fortify the front of her den.

He documented her powers of imitation as she changed the color and texture of her skin to disguise herself and escape from predators.

He witnessed the regrowth of an arm severed by ever-threatening sharks. Most importantly, he bonded with the creature, who learned to trust him and welcome his visits.

Surfacing Heart to Heart

The highlight of the documentary? The octopus embraced his chest as the two took a dramatic ride to the surface.

---

[811] Tiffany Duong, "'My Octopus Teacher' Stuns Audiences, Reinforces Power of Nature," EcoWatch, September 24, 2020, https://www.ecowatch.com/my-octopus-teacher-movie-2647785692.html.

This story of a human relationship with a mollusk won a well-deserved Academy Award. It demonstrated that creatures from alien worlds—one an Earthbound hominid, the other a relative of the clam—could use their God-given powers of comprehension to bond and befriend each other.

Landfall on Mars

While the octopus might be described as Earth's alien, NASA continues to search for remnants of alien life on Mars.

United States NASA rovers have made landfall on the planet Mars on six separate occasions.[812] The goal? To search for evidence that water once existed on Mars and to answer one of humanity's most fundamental questions: has life ever existed on another planet?

A 6th Mars rover explores the red planet, looking for
signs of past water and even traces of life.

---

[812] Dave Williams, "Chronology of Mars Exploration," NASA, last updated April 19, 2024, https://nssdc.gsfc.nasa.gov/planetary/chronology_mars.html.

Perseverance, the latest rover as of 2024, is the product of almost three decades of research involving three thousand NASA employees and more than four thousand non-government contractors from various companies and countries.[813] Together, they have assured that Perseverance has sufficient mobility to examine a large territory and to reposition itself should it encounter harsh atmospheric conditions.

NASA launched Perseverance on July 30, 2020, from Cape Canaveral, Florida. This newest rover contains the following part:[814]

1. A graphite fiber body to protect the rover's vital organs
2. A computer brain to process information
3. Temperature controls, including heaters and layers of insulation
4. A neck and head to provide a human-scale view
5. Eyes and ears in the form of cameras and instruments
6. An arm and "hand" to collect rock samples
7. Wheels and legs for mobility
8. Batteries for power
9. Communication antennas for "speaking" and "listening"

Among other items, Perseverance also carried the following additional equipment:

1. An advanced camera system to study surface minerals
2. A sensor to measure temperature, wind speed and direction, pressure, humidity, and dust
3. Experimental equipment to test the production of oxygen for future human missions to the planet

---

[813] "Mars 2020: Perseverance Rover," NASA, last updated August 5, 2024, https://mars.nasa.gov/mars2020/.
[814] "Mars 2020: Perseverance Rover."

4.   Even a helicopter named Ingenuity attached to its belly

Perseverance is one of the most amazing tools for exploration built by humankind. But the early reports are not encouraging. Perhaps covered by water in the distant past, Mars is now a barren rock with no active oceans or kelp forest.

The octopus is likely to remain our closest contact with a visitor from an "alien" world.

Theories of Consciousness

The relationship of mind and brain presents one of the greatest challenges to human understanding. Different brain regions process different sensory information at slightly different times, yet their output is bound together into a unified whole. How are the distributed components of consciousness bound into a seamless movie? To express it another way: where is the theater of mind?

Some of the brightest scientists and philosophers in the world have attempted to answer this question.

Integrated Information Theory (IIT)

In 2004, neuroscientist Giulio Tononi proposed a popular theory known as "integrated information theory" (IIT) that distributes the theater across the entire brain.[815] Consciousness, this theory proposes, is a generalized brain phenomenon. It arises from the brain's ability to integrate complex information, thus, the theory's name.

To test this concept, IIT researchers have delivered a pulse of magnetic energy into a patient's skull to induce a chain reaction across the cortex. A network of EEG sensors positioned outside the

---

[815] Giulio Tononi, "Integrated information theory," Scholarpedia, last modified June 23, 2015, 17:15, http://www.scholarpedia.org/article/ Integrated_information_theory.

head recorded the subsequent electrical signals unfolding over time. Their conclusion? EEG patterns paralleled the level of the patient's consciousness.

In unconscious patients, many in a vegetative state, the rhythms were waxing and waning. In conscious patients, the magnetic stimulation ignited a storm of activity across different regions of the brain—a sign, researchers believe, that neuronal networks across the brain were integrating information.[816]

Proponents of IIT also propose that consciousness is an inherent attribute of matter. Fundamental particles such as electrons may contain an element of consciousness, an idea called "panpsychism." According to this worldview, the universe is overflowing with conscious items.[817] Thermostats might have some level of conscious experience.

IIT has its critics. The influential philosopher John Searle complained that it depends on the misappropriation of the concept of information in his review of neuroscientist Christof Koch's book, *Consciousness*, about ITT: "[Koch] is not saying that information causes consciousness; he is saying that certain information just is consciousness, and because information is everywhere, consciousness is everywhere.... The view is incoherent."[818]

A tougher challenge remains. IIT fails to address the "hard problem." How would the integration of information across the brain explain the inner quality of experience, the taste of caramel

---

[816] Marcello Massimini and Giulio Tononi, *Sizing up Consciousness: Towards an Objective Measure of the Capacity for Experience*, trans. Frances Andersen (Oxford: Oxford University Press, 2018).

[817] John Horgan, "Can Integrated Information Theory Explain Consciousness? A Radical New Solution to the Mind–Body Problem Poses Problems of its Own," *Scientific American*, December 1, 2015, https://www.scientificamerican.com/blog/cross-check/can-integrated-information-theory-explain-consciousness/.

[818] Horgan, "Can Integrated Information Theory Explain Consciousness?"

or the smell of a rose? Advocates of IIT continue to press their case, but other theories compete for attention.

Global Neuronal Workplace Theory

Another prominent theory, the "global workspace theory," believes that key brain structures—a privileged subset of neural structures—create the theater of conscious experience.[819] Consciousness happens when certain parts of the brain broadcast to other parts, as though a teacher was summarizing a class discussion by posting key points on a whiteboard.

IIT involves the flow of complex information across the brain without depending on privileged brain structures. Global workspace theory postulates that privileged structures are real, that an exclusive group of collaborating brain centers posts information, creating consciousness as that information becomes accessible to other areas of the brain.[820]

EEG recordings have shown that an unconscious perception only produces isolated waves in the primary sensory cortex with no effect on other brain regions. When a stimulus reaches consciousness, waves in the primary sensory areas give way to "neural ignition." Sustained waves flow to distant areas of the prefrontal and parietal cortices, areas assigned the challenge of "association." Researchers suggest this broadcasting across the brain makes information available to other subsystems involved in the process of

---

[819] Bernard J. Baars, "Global Workspace Theory of Consciousness: Toward a Cognitive Neuroscience of Human Experience," *Progress in Brain Research* 150 (2005): 45–53, https://doi.org/10.1016/S0079-6123(05)50004-9.

[820] Rita Carter, *Exploring Consciousness* (Berkeley: University of California Press, 2002), 116.

making decisions, consolidating memories, and producing a stream of thought.[821]

Information isolated to a single sensorimotor system might be useful for something routine such as fast typing, but "subconscious" activities rarely pierce the veil of conscious thought. You become aware of information written onto this hypothetical "whiteboard" when it becomes available to a host of brain regions—to the brain's modules involved in language, planning, memory, etcetera. But the whiteboard has limited space. We can only access one conscious scene at any moment.[822]

Does the Brain's Electromagnetic Field Produce Consciousness?

Yet another plausible theory proposes that the source of conscious experience resides in the electromagnetic (EM) fields of the brain, not in neurons themselves.[823] We know that the brain is an electrochemical organ; its neurons are constantly producing electrical signals. Technicians can document the electric flow by attaching electrodes to the scalp and recording an EEG. Proponents of the EM field theory of consciousness propose that EM fields capture and consolidate neuronal output, creating the movie of the mind. If neurons fire randomly, they generate waves with peaks and troughs that cancel each other, creating a flatline field. But when neurons

---

[821] Gustavo Deco, Diego Vidaurre, and Morten L. Kringelbach, "Revisiting the Global Workspace Orchestrating the Hierarchical Organization of the Human Brain," *Nature Human Behaviour* 5 (2021): 497–511, https://doi.org/10.1038/s41562-020-01003-6.

[822] Lachlan Kent and Marc Wittmann, "Time Consciousness: the Missing Link in Theories of Consciousness," *Neuroscience of Consciousness* 2021, no. 2 (2021): article niab011, https://doi.org/10.1093/nc/niab011.

[823] Johnjoe McFadden, "Integrating Information in the Brain's EM Field: The CEMI Field Theory of Consciousness," *Neuroscience of Consciousness* 2020, no. 1 (2020): article niaa016, https://doi.org/10.1093/nc/niaa016.

fire in synchrony, peaks and troughs reinforce one another, creating a consolidated image on the conscious EM field.[824]

This model considers the brain a TV broadcasting station, and the brain's EM field a receiver. There is something curiously satisfying about this theory, because more than any other idea, it fits the common feeling that we are viewing a moving picture of the mind.

In the early 1900s, William James proposed a similar analogy: the brain is both a receiver and a transmitter for consciousness.[825] Many theorists would agree with him today. A material brain may do the "tuning," while the EM fields encircling the brain provide the "receiving"—producing the multimedia movie called the mind.

This theory of the brain as a receiver nicely addresses the "binding problem," the ability of our mind to integrate information across "time, space, attributes, and ideas."[826] Our consciousness combines different elements of a visual scene—its motion, sound, colors, etcetera. Widely different regions of the brain process a conscious scene's components, and yet somehow, those bits and pieces end up neatly assembled into a unitary conscious experience. We can look to our television sets for a similar, if not identical, process.

Other Theories of Quantum Mind

Quantum mechanics is a recurrent theme in any effort to understand the flow of human consciousness. Who can doubt the importance of superposition (one wave function contains the potential for a particle to materialize at innumerable locations) and entanglement

---

[824] McFadden, "Integrating Information in the Brain's EM Field."

[825] John R. Shook, *The Essential William James* (Amherst, NY: Prometheus Books, 2011).

[826] Anne Treisman, "Solutions to the Binding Problem: Progress through Controversy and Convergence," *Neuron* 24, no.1 (September 1999): 105–125, https://doi.org/10.1016/S0896-6273(00)80826-0.

(particles can pair up in teams). Without these quantum wonders, there would be no nuclear fusion in the stars. Who can help but be intrigued by carbon's perfect quantum leap during stellar fusion, nudging out beryllium-8 to maximize the production of carbon-12.

Quantum biology is now a respected field.[827] During photosynthesis, plants use quantum superposition to capture every photon of light.[828] Migratory birds use a "quantum compass" to guide their navigation around the globe.[829] Other scientists believe that quantum vibrations form the foundation for the human sense of smell.[830]

The role of quantum mechanics in consciousness has been a topic of never-ending speculation. "Quantum mind" is widely derided as "mystical woo," and the mere mention of "quantum consciousness" evokes images of New Age gurus, astrology, and psychics.[831] But our minds perform feats that stump digital computing. Our flow of thought and speech and our ability to recall long-forgotten memories feel like superpositions collapsing to provide the best answer. What if our conscious minds are somehow entangled with the quantum world?

---

[827] Johnjoe McFadden and Jim Al-Khalili, *Life on the Edge: The Coming of Age of Quantum Biology* (New York: Broadway Books, 2016).

[828] Artur Braun, *Quantum Electrodynamics of Photosynthesis* (Berlin: De Gruyter, 2020).

[829] Hamish G. Hiscock et al., "The Quantum Needle of the Avian Magnetic Compass," *Proceedings of the National Academy of Science* 113, no. 17 (April 2016): 4634–4639, https://doi.org/10.1073/pnas.1600341113.

[830] Chandler Burr, *The Emperor of Scent: A Story of Perfume, Obsession, and the Last Mystery of the Senses* (New York: Random House, 2003).

[831] Massimo Pigliucci, *Nonsense on Stilts: How to Tell Science from Bunk* (Chicago: University of Chicago Press, 2018).

What Structure in the Brain Could Serve as a Quantum Processor?

In "Orch OR theory," Stuart Hameroff and Roger Penrose have proposed that microtubules within neurons are the site of quantum processing.[832] Critics reject that proposal out of hand, confident that the brain's heat would quickly destroy the superposition required for quantum calculations inside the neuron. A single photon colliding with a qubit, they remind us, can make the entire system fall apart, eliminating entanglement and wiping out the system's quantum properties.

Yet another theory suggests that consciousness reflects the superposition of electromagnetic fields within the neocortex, delivering information at the speed of light to a central processor called the "thalamus," an egg-shaped mass of grey matter in the forebrain receiving sensory information from all over the body.[833] In this model, the thalamus represents the gathering place for the components of the mind.

Sorting Through the Theories

What conclusions can we reach about these theories of consciousness? Information is surely involved, and it is reasonable to believe that it must be complex and integrated to produce a meaningful conscious frame (integrated information theory).

---

[832] Stuart Hameroff and Roger Penrose, "Consciousness in the Universe: A Review of the 'Orch OR' Theory," *Physics of Life Reviews* 11, no. 1 (March 2014): 39–78, https://doi.org/10.1016/j.plrev.2013.08.002.

[833] Lawrence M. Ward and Ramón Guevara, "Qualia and Phenomenal Consciousness Arise from the Information Structure of an Electromagnetic Field in the Brain," *Frontiers in Human Neuroscience* 16 (2022): article 874241.

Creating a whiteboard through a privileged set of neural structures could focus that information and facilitate its retrieval (global workplace theory).

And the rapid flow of thought and emotion feels like a quantum mechanical process: it makes sense that quantum mechanisms might collect and organize the chaotic output of our busy brains (quantum field theory).

But a leading philosopher of mind, David Chalmers, would remind us that we have failed to address the "hard problem" of consciousness—the qualia of subjective experience. Wherever the theatre of the mind exists in our brain, how does it produce the irritation of an itch, or the softness of fur, or a stream of thought and emotion?[834]

Two Leading Theories Face-Off

IIT and global workplace theory have squared off on an "adversarial collaboration" in a series of experiments begun in 2020,[835] pursuant to a twenty-five-year-old bet. Philosopher David Chalmers of "hard problem" fame and renowned neuroscientist Christoff Koch, proponent of IIT, wagered that the "neural correlates" of consciousness would be identified (Koch) or would remain mysterious (Chalmers). Chalmers was skeptical that either ITT or Global Workplace could explain the deep (and hard) mystery of subjective experience.

IIT suggests that conscious activity arises in the posterior portion of the brain, while global theory places the action in the prefrontal cortex.

---

[834] David J. Chalmers, *Philosophy of Mind: Classical and Contemporary Readings*, 2nd ed. (Oxford: Oxford University Press, 2021).
[835] "Thousands of Species of Animals Probably Have Consciousness," *Economist*, June 28, 2023, https://www.economist.com/science-and-technology/2023/06/28/thousands-of-species-of-animals-likely-have-consciousness.

Over two hundred volunteers, some in a functional MRI scanner, others evaluated by an advanced form of EEG called magnetoencephalography, looked at a series faces, letters, and shapes and pushed a button when they reached a conscious experience. Would the clearest signal come from the prefrontal cortex as privileged centers informed the rest of the brain, or would the conscious experience arise in the rear of the brain where neurons have the most complex connections? Chalmer's bet: neither portion of the brain would correlate with the conscious experience.

Chalmers emerged the winner, as summarized by Anil Seth, neuroscientist at the University of Sussex in England: "The current experiment is enough to show that neither theory is presently sufficient."[836]

This comes as no surprise. See the dual substance theory of primordial mind mapping to the connectome of the brain in supplement 27.

Is Mind "What the Brain Does?"

The take-home point of this supplement should be obvious: subjective conscious experience remains a colossal mystery. Each of these theories assumes that consciousness is a product of the brain. Under the model of "substance dualism," we see consciousness in another light. God dispersed his mind into creation and created human beings in His image. Whatever the role of the brain, it is subservient to a primordial substance, nothing less than the mind of our Creator.

---

[836] Carl Zimmer, "2 Leading Theories of Consciousness Square Off," *New York Times*, July 1, 2023, www.nytimes.com/2023/07/01/science/consciousness-theories.html.

Math and Science

As a patent office worker, Albert Einstein mentally rode a beam of light and revolutionized the field of physics.[837] Energy and matter are different forms of the same thing.[838] Distance and time are relative to the frame of motion.[839] Gravity is less a cosmic force and more a reflection of the degree to which energy and matter curve the space-time continuum.[840]

After leaving the Patent Office in 1909, Einstein joined the faculty at the University of Zurich and spent the next decade searching for mathematics that would support his general theory of relativity. He found a perfect fit in the form of Riemannian geometry, a mathematical analysis of curved surfaces developed in the 1850s.[841]

In the last few decades of his life, Einstein hoped to develop a "theory of everything" that would connect relativity with electromagnetism. In the absence of experimental data, his mathematical efforts proved fruitless. Einstein's life and career provide a perfect

---

[837] Walter Isaacson, *Einstein: His Life and Universe* (New York: Simon and Schuster, 2007).

[838] Albert Einstein, "Does the Inertia of a Body Depend Upon Its Energy Content?" *Annelen der Physik*, September 27, 1905.

[839] Richard Wolfson, *Simply Einstein: Relativity Demystified* (New York: W. W. Norton and Company, 2003), 18.

[840] Albert Einstein, *Relativity: The Special and the General Theory - 100th Anniversary Edition* (Princeton: Princeton University Press, 2015).

[841] Luther Pfahler Eisenhart, *Riemannian Geometry* (Princeton, NJ: Princeton University Press, 1997).

introduction to the complex relationship between physics and mathematics. Does math describe reality, or create it?

Math as a Fundamental Searchlight

In ancient Greece, Pythagoreans worshipped numbers and believed that geometry would solve the riddles of the world.[842] Plato believed in an eternal mathematical landscape populated by ideal forms.[843]

In 1928, Paul Dirac encountered the equivalent of an ideal form in a beautiful equation, suggesting that every fundamental particle has of anti-matter twin.[844] Experiments confirmed the existence of anti-electrons,[845] leading one observer to declare that Dirac "was the first person to glimpse the other half of the early universe, entirely through the power of reason."[846]

Anti-matter particle pairing also suggests that space-time could never be empty but would be constantly seething with particles and anti-particles flickering in and out of existence.[847]

In 1967, Roger Penrose used the power of his mind's eye to develop the mathematical description of black holes.[848] Not only did Penrose provide a description of theoretical black holes, but he also proved they exist throughout the universe. Cosmology now

[842] W. K. C. Guthrie, *A History of Greek Philosophy, Volume 1: The Earlier Presocratics and the Pythagoreans* (Cambridge: Cambridge University Press, 1979), 173.

[843] Plato, *Great Dialogues of Plato*, trans. W. H. D. Rouse (New York: Signet, 2015), xvi.

[844] Graham Farmelo, *The Strangest Man: The Hidden Life of Paul Dirac, Mystic of the Atom* (New York: Basic Books, 2011).

[845] Farmelo, *The Strangest Man*, 434.

[846] Farmelo, *The Strangest Man*, Farmelo, 2.

[847] Gordon Kane, "Are Virtual Particles Constantly Popping In and Out of Existence? Or are they merely a mathematical bookkeeping device for quantum mechanics?" *Scientific American*, October 9, 2006, https://www.scientificamerican.com/article/are-virtual-particles-real/.

[848] Roger Penrose, *The Road to Reality: A Complete Guide to the Laws of the Universe* (New York: Vintage Books, 2004), 691.

understands that a black hole lives at the center of every galaxy, and recent theory suggests that black holes may provide the dark energy that drives the expansion of the universe.[849]

Observation and Experimentation: Guides to Mathematical Theory

Dirac and Penrose used beautiful equations and the power of their mind's eye to unlock secrets of the universe. But sheer mathematical reasoning proved to have its limits. Galileo's observations of falling bodies and Newton's analysis of gravity gave birth to another concept. Scientific observation and experimentation could stimulate the search for mathematical explanations.

Analysis of a mathematical equation can lead to a new scientific discovery. Peter Higgs discovered the mass-encoding field known by his name by analyzing an equation in 1964.
On other occasions, observation and data produce scientific progress. Scattering experiments performed at the Stanford Linear Accelerator led to the discovery of quarks.
Math and experimentation are twin pillars and partners in the march of scientific process.

---

[849] Duncan Farrah et al., "Observational Evidence for Cosmological Coupling of Black Holes and its Implications for an Astrophysical Source of Dark Energy," *The Astrophysical Journal Letters* 944 (2023): L31, https://doi.org/10.3847/2041-8213/acb704.

Newton was the first scientist to explain his work with simple and elegant equations.[850] Gravity, his math reveals, varies with the relative mass of material objects and inversely with the distance between them.

Physicists in the 1960s smashed the atom and developed the quantum field theory of the standard model.[851] We should be reminded yet again that the standard model needed an explanation for mass. The Higgs field that encodes mass and gives our universe physical structure was discovered in a mathematical equation.[852]

In 2012, collider experiments at CERN in Geneva, Switzerland demonstrated the reality of the Higgs boson. From experiment to math, from math to experiment, the mystery remains as described by an avid science reader, even citing the title of physicist Eugene Wigner's important paper: "The unifying insight across almost every major discovery is the 'unreasonable effectiveness of mathematics in the natural world.'"[853]

Betrayed by Mathematics

Not all beautiful math has provided a reliable guiding light. String theory has suggested that our universe has hidden higher dimensions, some of which undergo compactification to control the vibrations of strings.[854] Those compactified dimensions play

---

[850] James Gleick, *Isaac Newton* (New York: Vintage Books, 2003).

[851] Robert Oerter, *The Theory of Almost Everything: The Standard Model, the Unsung Triumph of Modern Physics* (New York: Penguin Group, 2006).

[852] Peter W. Higgs, "Broken Symmetries and the Masses of Gauge Bosons," *Physics Review Letters* 13, no. 16 (October 1964): 508, https://doi.org/10.1103/PhysRevLett.13.508.

[853] Eli Katz, review of *The Universe Speaks in Numbers: How Modern Maths Reveals Nature's Deepest Secrets*, by Graham Farmelo, Amazon, May 8, 2019, https://www.amazon.co.uk/gp/aw/review/B07PLK7R4N/R11F5LJ8JFEMGU.

[854] Sabine Hossenfelder, *Lost in Math: How Beauty Leads Physics Astray* (New York: Basic Books, 2018).

the strings like musical notes and produce the different particles of the standard model. String theory and its companion theory, supersymmetry, can be used to describe a near-infinite variety of universes.[855] Enamored with the beauty of this math, numerous authors have joined the string theory band wagon, producing the following list of books:

- *Parallel Worlds* by Michio Kaku
- *The Cosmic Landscape* by Leonard Susskind
- *Many Worlds in One* by Alex Vilenkin
- *The Goldilocks Enigma* by Paul Davies
- *In Search of the Multiverse* by John Gribbin
- *From Eternity to Here* by Sean M. Carroll
- *The Grand Design* by Stephen Hawking
- *The Hidden Reality* by Brian Greene
- *Edge of the Universe* by Paul Halpern

In *The Mind of God*, Paul Davies expresses his concern with the concept of a multiverse.[856] Why postulate an infinite succession of universes to explain the exquisite fine-tuning of one universe, the one in which we live? Occam's razor demands a simpler explanation, and Davies prefers the hypothesis of design. Without explicitly saying so, Davies would attribute the tuning of our world to divine design rather than the randomizing effect of unseen and unknowable universes.

In *Our Mathematical Universe*, Max Tegmark takes the final step, declaring that math not only describes the universe but also

---

[855] Hossenfelder, *Lost in Math*, 100.
[856] Paul Davies, *The Mind of God: The Scientific Basis for a Rational World* (New York: Simon and Schuster, 1993), 190.

creates it.[857] The creation is prolific, producing four different levels of multiverse, according to Tegmark. Electrons and human beings, in his account, are more than particles or flesh; they are mathematical structures.[858] It is fair to say that many scientists, Paul Davies probably included, remain unconvinced.

The beauty and effectiveness of mathematics remains a scientific and philosophical puzzle. In *The Universe Speaks in Numbers*, Graham Farmelo may have the last word, stating, "Physicists have not one but two ways of improving their fundamental understanding of how nature works: by collecting data from experiments and by discovering the mathematics that best describes the underlying order of the cosmos. The universe is whispering its secrets to us, in stereo."[859]

There is yet another explanation. God is real. His thoughts are mathematical, and He is more than prepared to answer the challenge from Stephen Hawking: "What is it that breathes fire into the equations and makes a universe for them to describe?"[860]

---

[857] Max Tegmark, *Our Mathematical Universe: My Quest for the Ultimate Nature of Reality* (New York: Vintage Books, 2014).

[858] Tegmark, *Our Mathematical Universe*, 6.

[859] Graham Farmelo, *The Universe Speaks in Numbers: How Modern Math Reveals Nature's Deepest Secrets* (New York: Basic Books, 2019).

[860] Hawking, 1988. A Brief History of Time. Chapter 11.

Altered Memory

Decades of research have examined the proposal that the mind can extend beyond the brain to connect with other minds. Should telepathy be real, it would represent a significant challenge to "physicalism," the material worldview. But we encounter the wonders of human consciousness in everyday life: the ability to create memories that tell the stories of our lives.

The Movie of Our Lives

What becomes a favorite pastime when visiting old friends or aging loved ones? We recall shared memories from childhood and reflect on the journey of life. When a great-nephew visits a hundred-year-old aunt, he'll ask the following questions: "How did you live without indoor plumbing or running water? How did you survive summers without air conditioning? What was it like with no television, even no internet?"

She'll reply by relaying favorite memories. Without indoor plumbing, sanitation trucks arrived to evacuate outhouses, and children abandoned the streets. Families gathered for supper and talked to each other without interruption, eating food from the farm, prepared by family members.

Neuroplastic Brain

As we form memories, the brain changes its physical structure. Synapses, the connection between neurons, grow stronger as neuronal networks stitch memories across the brain.[861] Neuroscientists still struggle to understand the neurobiology of memory formation, but observers agree that memories serve a higher purpose:

They enrich our lives in the here and now and support our dreams of the future.

Key Definition: Memory

Synaptic strengthening creates memories, stitching networks across the brain that allow us to recall past experiences and conceptualize possibilities for the future.

1. *Episodic memories are created in the hippocampus in the brain's temporal lobe.*
2. *Information stored in the hippocampus can be transferred to the neocortex during sleep.*
3. *The amygdala, an almond-shaped structure in the temporal lobe, attaches emotional significance to memories, making them readily available for recall.*

---

[861]   John Lisman et al., "Memory Formation Depends on Both Synapse-Specific Modifications of Synaptic Strength and Cell-Specific Increases in Excitability," *Nature Neuroscience* 21 (2018): 309–314, https://doi.org/10.1038/s41593-018-0076-6.

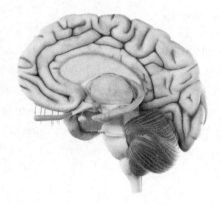

The amygdala recognizes danger and fear. The hippocampus processes memories during our sleep. The cerebellum stores memories of movement such as riding a bike. The prefrontal cortex analyzes the significance of memories. The entire brain is a memory machine.

Memories define who we are and tell the story of our lives. As Dr. James McGaugh, an American neurobiologist working in learning and memory expressed it, "Memory is our bridge to the future."[862] In some cases, that bridge seems enormously broad. In others, surgery or trauma blows the bridge away. And sometimes people have an extraordinary ability to recall memories that to most people would remain buried and forgotten.

[862] Linda Rodrigues McRobbie, "Total Recall: The People Who Never Forget," *The Guardian*, February 8, 2017, https://www.theguardian.com/science/2017/feb/08/total-recall-the-people-who-never-forget.

Highly Superior Autobiographical Memory (HSAM)

Daniel McCartney, a fifty-seven-year-old blind man living in Ohio, could remember the day of the week, the weather, what he was doing, and where he was on any date going back to January 1, 1827, when he was nine years and four months old.[863]

Jill Price, a thirty-four year-old woman, can recall details from every day of her life. The third time she drove a car was on January 10, 1981, a Saturday, nearly twenty years earlier. She first heard the Rick Springfield song "Jessie's Girl" on March 7, 1981, while she was driving in a car with her mother, who was yelling at her. She was sixteen years and two months old.[864]

Some sixty people in the world, perhaps more, have "highly superior autobiographical memory," or HSAM. They can remember most of the days of their life as clearly as the rest of us remember yesterday. A morning talk show host demonstrated her HSAM by asking her cohost to name a random year. Within seconds, the host with HASM remembered the day of the week for every holiday that particular year. HSAM, she says, is like time travel.[865]

Individuals with this syndrome have noticeable increases in an area of the brain involved in recalling emotional memories. Brain scans also demonstrate increases in the uncinate fasciculus, a bridge between the frontal and temporal cortex involved in episodic memory retention. A scientific article described people with HSAM as

---

[863] W. D. Henkle, "Remarkable Cases of Memory," *Journal of Speculative Philosophy* 5, no. 1 (January 1871): 6–26, https://www.jstor.org/stable/25665736.

[864] McRobbie, "Total Recall."

[865] Aurora K. R. LePort et al., "Highly Superior Autobiographical Memory: Quality and Quantity of Retention Over Time," *Frontiers Psychology* 6 (2016): article 2017, https://doi.org/10.3389/fpsyg.2015.02017.

both the "warden and prisoner" of their memories."[866] They have lost their ability to forget!

In sharp contrast, individuals can lose their ability to remember. In 1953, a twenty-seven-year-old man had a "bilateral medial temporal lobe resection," or lobectomy, to stop life-threatening seizures. Surgeons destroyed his hippocampus and much of his amygdala, and the surgery left him completely unable to form new memories. He could remember life before the surgery and learn new motor skills, but every day he reintroduced himself to the researcher who counseled him, because he never remembered having met her in the past.[867]

A forty-six-year-old man slipped on the bathroom floor at work and suffered a blow to the back of his head, immediately losing the ability to remember things from his past. Over the subsequent year, he had to remeet family and friends, relearn his life story, and rebuild his self-identity. His wife of twenty-five years helped by showing him photographs of their life arranged in chronological order.

A brain scan showed a loss of blood flow to his right temporal lobe. The patient himself described his condition: "If my life was a keyboard, someone pressed the delete button, and all my memory is gone."[868]

Telepathy would involve the connection of our mind with another mind. Memories do even more. They connect our mind today with the mind of yesterday, and hopefully, with the minds of tomorrows.

---

[866] McRobbie, "Total Recall."
[867] McRobbie, "Total Recall."
[868] "Man with Amnesia Lost 46 Years in Workplace Slip," ABC News, April 16, 2010, https://abcnews.go.com/Nightline/amnesia-man-hits-head-loses-memories/story?id=10396719.

Children Who Remember Past Lives

In chapter 22, we explored the ability of the mind to function despite a severely compromised brain. We reviewed reports of children with advanced hydrocephalus and compressed brain tissue who possess normal intelligence and live normal lives. We also visited the phenomenon of "terminal lucidity," patients with advanced dementia who seem to regain normal mental function in the last few days or hours of their life. Are their minds escaping a sick and dying brain?

In this supplement, we approach this issue from another direction. Can someone who has "escaped the surly bonds of Earth" live yet another life?[869]

The Division of Perceptual Studies

In 1958, Dr. Ian Stevenson, chairman of the Department of Psychiatry at the University of Virginia, published a paper titled "The Evidence for Survival from Claimed Memories of Former

---

[869] Tucker, Return to Life, 88.

Incarnations."[870] His review focused on a collection of forty-four cases from different countries suggesting the possibility of reincarnation. These cases typically involved young children with memories of a past life. Impressed by these reports, Stevenson traveled to India and investigated twenty-five cases, soon followed by a trip to Ceylon (now Sri Lanka) to review an additional five cases.

Interested in this work, Chester Carlson, the inventor of Xerox, provided funding for the Division of Perceptual Studies at the University of Virginia,[871] and Dr. Ian Stevenson resigned his position as chairman of psychiatry to study past-life memories for the remainder of his career. In 1966, he published his personal research on the topic, *Twenty Cases Suggestive of Reincarnation*, in which he detailed reports of cases from India, Ceylon, Brazil, and Lebanon.[872]

In 1977, the *Journal of Nervous and Mental Disease* devoted most of one issue to Dr. Stevenson's work, and researchers began to pursue the topic worldwide.[873]

Studies in the US

In 1968, Dr. Jim Tucker left the private practice of psychiatry to join Dr. Stevenson and pursue past-life memories in the United

---

[870] Ian Stevenson, "The Evidence for Survival from Claimed Memories of Former Incarnations," *Journal of the American Society for Psychical Research* 54, no. 2 (April 1960): 51–71, https://med.virginia.edu/perceptual-studies/wp-content/uploads/sites/360/2016/12/STE1.pdf.

[871] "About," Division of Perceptual Studies, University of Virginia, date accessed XXX, https://med.virginia.edu/perceptual-studies/who-we-are/.

[872] Ian Stevenson, *Twenty Cases Suggestive of Reincarnation* (Charlottesville: University of Virginia, 1966).

[873] *Journal of Nervous and Mental Disease* 165, no. 3 (September 1977).

States. In numerous publications, including his well-received 2013 book *Return to Life*, he has reported the following findings.[874]

1. Almost two-thirds of the people who reported having a past-life experience state that their previous life ended due to an unnatural cause, most commonly an accident, suicide, or another violent act.

2. In one out of five cases, unusual birthmarks matched the fatal wounds of the former individual.

3. The time between life and possible rebirth averaged sixteen months.

4. In almost three-quarters of cases, the "previous personality" died at a relatively young age, with a quarter dying before their fifteenth birthday.

5. The child remembering a past life was born, on average, four-and-a-half years after the death of the "remembered individual."

Given the controversial nature of the topic, researchers in the field have carefully reviewed the details of each child's story, attempting to exclude mundane sources of information, such as parental embellishment or childhood fantasizing. As a reality check, researchers ask their young subjects to identify pictures relevant to their past life from a random assortment of images: does one of these pictures match your past-life house? Which photo represents your previous wife?

Two cases summarized in *Return to Life* present compelling past-life stories.

---

[874] Jim B. Tucker, *Return to Life: Extraordinary Cases of Children who Remember Past Lives* (New York: St. Martin's Press, 2015).

A Hollywood Talent Agent

At age three, Ryan Hammons declared to his parents that he'd "been" someone else before.[875] He remembered living in Hollywood with three sons in a house with a swimming pool. Ryan often pretended to be filming and snapping clapboards and knew tap dance routines he'd never been taught. Recalling that his favorite Hollywood restaurant had been Chinese, Ryan could use chopsticks without any prior instruction. He recalled that his favorite drink had been "Tru Ade," an orange soda that went out of production before Ryan was born.

Ryan and his mother reviewed library books with Hollywood as a topic. In one of those books, a black-and-white photo showed a group of people from the film *Night After Night*. Ryan excitedly pointed to one of the men, telling his mother, "That's me. That's who I was."[876] Eventually, the individual in the photo was identified as a Hollywood agent named Marty Martyn.

Beginning at Ryan's age five, Ryan's mother recorded his memories in a journal. When Marty Martyn's identity was finally known, it was determined that Ryan had made forty-seven correct statements about Marty Martyn's life.

Ryan seemed to make a mistake when he asserted that he (Marty Martyn) had died when he was sixty-one, because Martyn's death certificate gave his birth date as 1905, which would mean that he died at age fifty-nine. Subsequent research showed that, in fact, Martyn was born in 1903, making Ryan's memory spot on.

---

[875]   Tucker, *Return to Life*, 88.
[876]   Jake Whitman and Cynthia McFadden, "'Return to Life': How Some Children Have Memories of Reincarnation," Today, March 17, 2015, https://www.today.com/news/return-life-how-some-children-have-memories-reincarnation-t8986.

As is typical of "past life" children, the memories of Hollywood faded as Ryan matured and developed memories of his own.

A WWII Fighter Pilot

When James Leininger was only two-and-a-half years old, he began to experience past-life memories of a Lt. James McCready Huston, a World War II fighter pilot from Uniontown, Pennsylvania, who died in an air battle over Iwo Jima.[877] He informed his bemused parents that he had been assigned to an aircraft carrier called the *Natoma* and had flown a plane called a Corsair. When his parents served meatloaf for lunch, which he had never eaten before that day, he said he hadn't meatloaf since he was on the carrier.

When his mother pointed out "a bomb on the bottom" of a WWII model airplane, two-and-a-half-year-old James quickly corrected her, saying that it was drop tank. Drop tanks were attached to Corsairs to provide extra fuel.

James had nightmares about being shot down by a Japanese plane with a red sun painted on the side. During recurrent dreams, he screamed at the top of his voice, "Airplane crash, on fire, can't get out, help!"

James's father decided to do some research of his own. He discovered there had been a small escort carrier called the *Natoma Bay* involved in the Battle of Iwo Jima. Further research proved there had indeed been a pilot named James Huston, and his plane had indeed been hit in the engine by Japanese fire on March 3, 1945.

In a further twist to the story, Jim Tucker and James's parents visited Huston's sister, Anne Barron, now eighty-seven years old.

---

[877] Tucker, *Return to Life*, 64.

After listening to little James's story, she declared, "He knows too many things; for some reason, he knows what happened."

In 2004, when James was six years old, his father took him to a reunion of veterans who had served on the *Natoma Bay*. After sixty years, James recognized one of his old mates, declaring to his parents with surprise, "They're so old."

Past-Life Memories Fade

The following is an important feature of "past live" cases: children usually "remember" their previous experiences when they're between the ages of two and seven. Memories begin to fade as the kids grow older. They inevitably become less invested in the past and more involved in their ongoing life.[878]

Potential Explanations

Although reincarnation is a prominent feature of Buddhist and Hindu religions and most past-life memories are reported from Asian countries, reincarnation is not a common belief in the Judeo-Christian tradition. There is little evidence that memories can be transmitted through DNA to a newborn infant brain. Is there an alternative explanation for these stunning reports?

A detailed life review is a prominent component of a near-death experience.[879] Experiencers describe revisiting their life in granular detail, often from the perspective of their impact on other people. If these stories are true, they suggest each of us leaves a detailed record of our life in the memory bank of the cosmos.

---

878 Tucker, *Return to Life*, 85.
879 Bruce Greyson, *After: A Doctor Explores What Near-Death Experiences Reveal about Life and Beyond* (New York: St. Martin Essentials, 2021).

Are two- to three-year-old children with newly developing brains absorbing the memory traces of an unrelated life until they develop memories of their own?

Whatever the correct explanation, these stories suggest that, yet again, the standard position of materialism, the "mind is what the brain does," is an impoverished view of the world.

Ancient and Modern Near-Death Reports

In chapter 23, we reviewed the mounting evidence for the reality of the near-death experience (NDE). Thousands of reports, more accumulating every day, suggest that the phenomenon is real and reproducible.

While the NDE might seem like a phenomenon of modern culture, the writings of the Apostle Paul suggest that the NDE is an ancient phenomenon. In letters approaching two thousand years old, he alludes to an experience very much compatible with a modern NDE.

Following his encounter with "the Lord's glory"[880] he spent at least three years in Arabia before launching his apostolic mission.[881] Apparently, Paul was as reluctant to speak of his experience as many modern survivors today are.

Did Paul have an out-of-body experience? He answers that question shyly, in the third person: "I know a man in Christ" taken up to heaven. "Whether it was in the body or out the body, I do not know—God knows."[882] Did Paul encounter an otherworldly light?

---

880   2 Corinthians 2:18 (NIV).
881   Galatians 1:17 (NIV).
882   2 Corinthians 112:2 (NIV).

Luke has Paul describe a heavenly light in three separate accounts in the book of Acts.

Did Paul experience a spiritual body? He coined the term.

Did Paul undergo a life review? He tells us in multiple letters to expect one.

Did he return from his experience with special knowledge and a sense of mission? He devoted the remaining years of his life to the Christian gospel.

Did his experience make him less afraid of death? He suffered beatings, stoning, shipwrecks, snake bites—but declared to the end of his life: "To live is Christ. To die is gain."[883]

Paul never gave a single coherent account of his experience in any single letter, as though he feared it might interfere with his singular focus on the story of Christ. But he scattered hints throughout his writings. Let's compare his brief allusions with reports from modern literature.

Out of Body

Modern Report

> "I could feel my spirit actually leaving my body. I saw and heard the conversations between my husband and doctors taking place outside my room, about forty feet away down a hallway."[884]

> "I found myself floating up toward the ceiling. I could see everyone around the bed very plainly, even my own body."[885]

---

[883] Philippians 1:21 (NIV).
[884] Jeffery Long and Paul Perry, *Evidence of the Afterlife: The Science of Near-Death Experiences* (New York: HarperOne, 2011), 6.
[885] Long and Perry, *Evidence of the Afterlife*, 25.

Paul

> 2 Corinthians, 12:2 (NIV): "I know a man in Christ who fourteen years ago was caught up to the third heaven. Whether it was in the body or out of the body I do not know—God knows."

Mystical Light

Modern NDE

"At first the light was blue. Then it transitioned to white. It was an opalescent white; it almost glowed, but did not shine. It was bright, but not intense bright, like glowing bright—pure bright. Pure but not in the usual sense of the word. Pure as in something you've never seen before or could ever describe or put into words."[886]

"The landscape was beautiful, blue skies, rolling hills, flowers. All was full of light, as if lit from within itself and emitting light, not reflecting it."[887]

"I am instantly drawn toward the light—I can feel its brightness, warmth, and love. As I get closer to it, I am absorbed by its brilliance and perfect love. Oh my God, I am the Light!"[888]

---

[886] Long and Perry, *Evidence of the Afterlife*, 10.
[887] Long and Perry, *Evidence of the Afterlife*, 14.
[888] Jeffery Long and Paul Perry, *God and the Afterlife: The Groundbreaking New Evidence of Near-Death Experience* (New York: HarperOne, 2017), 79.

"I was just in…a beautiful landscape setting of grass, lawns, and trees, and brilliant light."[889]

"The magnificent light I was experiencing…was brighter than the sun."[890]

Paul

Acts 9:3 (NIV): "As he neared Damascus on his journey, suddenly a light from heaven flashed around him."

Ephesians 5:8 (NIV). "Wake up, O Sleeper, rise from the dead, and Christ will shine on you."

Colossians 1:15 (NIV). "(Give) thanks to the Father, who has qualified you to share in the inheritance of the saints in the *kingdom of light.*"

Life Review

Modern NDE

"I saw my life flash before me shortly after I left my body.…I saw every important event that had ever happened in my life, from my first birthday to my first kiss to fights with my parents."[891]

"Next, he showed me my life review. Every second from birth until death, you will see and feel, and

[889] Petere Fenwick and Elizabeth Fenwick, *The Truth in the Light: An Investigation of Over 300 Near-Death Experiences* (London: White Crow Books, 2012), 73.
[890] John Burke, *Imagine Heaven: Near-Death Experiences, God's Promises, and the Exhilarating Future that Awaits You* (Ada, MI: Baker Books, 2015), 102.
[891] Long and Perry, *Evidence of the Afterlife*, 13.

[you will] experience your emotions and others that you hurt and feel their pain and emotions."[892]

"Everything I ever did, said, hated, helped, did not help, should have helped, was shown in front of me, the crowd of hundreds and everyone like [in] a movie."[893]

"My whole life was there, every instant of it.... Everyone and everything I had ever seen and everything that had ever happened was there."[894]

Paul

Romans 2:16 (NIV). "This will take place on the day when God judges people's secrets through Jesus Christ, as my gospel declares"

1 Corinthians 4:5 (NIV). "It is the Lord who judges me...wait until the Lord comes. He will bring to light what is hidden in darkness and will expose the motives of the heart."

2 Corinthians 5:10 (NIV). "For we must all appear before the judgment seat of Christ, so that each of us may receive what is due us for the things done while in the body, whether good or bad."

Romans 14:10 (NIV). "For we will all stand before God's judgment seat. It is written: 'As surely as I

---

[892] Long and Perry, *Evidence of the Afterlife*, 14.

[893] Long and Perry, *Evidence of the Afterlife*, 114.

[894] Kenneth Ring and Evelyn Elsaesser Valarino, *Lessons from the Light: What We Can Learn from Near-Death Experiences* (Newburyport, MA: Moment Point Press, 2006), 148.

live, says the Lord, every knee will bow before me;
every tongue will acknowledge God.' So then, each
of us will give an account of ourselves to God."

Encountering or Learning Special Knowledge

Modern NDE

"When I looked into his eyes, all the secrets of the
universe were revealed to me."[895]

"In that state of total recall, I became one with all
life, part of a collective consciousness, and I knew
everything."[896]

"I was like I was a computer and receiving an
endless download of knowledge. I asked questions
and received answers immediately."[897]

Paul

Ephesians 3:9 (NIV). "Surely you have heard
about the...mystery made known to me by
revelation, as I have already written briefly.... This
mystery is that through the gospel the gentiles are
heirs together with Israel, members together of
one body, and sharers together in the promise in
Christ Jesus."

---

[895] Long and Perry, *Evidence of the Afterlife*, 15.
[896] Long and Perry, *God and the Afterlife*, 124.
[897] Long and Perry, *God and the Afterlife*, 124.

2 Corinthians 12:2 (NIV). "And I know that this man was caught up to paradise and heard inexpressible things, things that no one is permitted to tell."

Spiritual Body

Modern NDE

"I still had a 'body,' but it was entirely different. I could see in three dimensions as if I had no body at all but was just a floating eyeball.... I could see all directions at once."[898]

"It's sort of a translucent, pearlescent, shimmery. A brilliance of light—just exploding."[899]

Paul

1 Corinthians 15:43 (NIV). "So will it be with the resurrection of the dead. The body that is sown is perishable, it is raised imperishable; it is sown in dishonor, it is raised in glory; it is sown in weakness, it is raised in power; it is sown a natural body, it is raised a spiritual body."

1 Corinthians 15:45 (NIV). "If there is a natural body, there is also a spiritual body."

---

[898] Long and Perry, *Evidence of the Afterlife*, 60.
[899] Burke, *Imagine Heaven*, 64.

Love as the Primordial Substance and the Power of Creation

Modern NDE

> "I knew that the being I met was composed of a substance I can only call 'love,' and that substance was a force or power, like electricity."[900]

> "I felt the presence of pure love.... God is love, we are love, and love creates all that is."[901]

> "The entire encounter was about God, the ultimate power of God, and God's forgiveness. The message was, 'Love is the greatest power in the universe.'"[902]

> "The first message I was given was that the most important thing in the universe is love: love is all that matters; we are all the same, and we are all love."[903]

Paul

> I Corinthians 13 (NIV). "If I speak in the tongues of men and of angels, but have not love, I am a noisy gong or a clanging cymbal. And if I have prophetic powers, and understand all mysteries and all knowledge, and if I have all faith, so as to remove mountains, but have not love, I am nothing. If I give away all I have, and if I deliver

---

[900] Long and Perry, *God and the Afterlife*, 53.
[901] Long and Perry, *God and the Afterlife*, 50.
[902] Long and Perry, *God and the Afterlife*, 50.
[903] Long and Perry, *God and the Afterlife*, 8.

up my body to be burned—but have not love,
I gain nothing.… So now faith, hope, and love
abide, these three; but the greatest of these is love."

Changed Life, No Fear of Death

Modern NDE

"I had always been terrified of death, of oblivion.
I no longer fear death."[904]

"I am no longer afraid of death. I know now in
my soul that there is so much more after life. I feel
that once I have learned what it is I am supposed
to learn or a task that I must complete, that I will
be rewarded with a life after death!"[905]

Paul

2 Corinthians 5:8 (NIV). "We are confident, yes,
well pleased rather to be absent from the body and
to be present with the Lord."

Philippians 1:21 (NIV). "For to me, to live is
Christ, and to die is gain."

More to the Story: Near-Death Experience

Skeptics repeatedly attempt to explain near-death experiences
as some malfunction of the brain. Near-death survivors, they claim,

---

[904] Long and Perry, *Evidence of the Afterlife*, 192.
[905] Long and Perry, *Evidence of the Afterlife*, 192.

must have had a temporal lobe seizure,[906] altered blood gas levels,[907] or memories of passing through the birth canal into a brightly lit obstetric suite.[908] Perhaps the NDE was simply the effect of a medication.[909] Surely, the presence of otherworldly light, an encounter with the predeceased, or an encounter with a god-like figure must reflect a neurobiological disorder or a psychological disease.

As previously noted in chapter 23, two aspects of the NDE argue against their origin in a hallucination or dream. First, all near-death stories reach a firm conclusion. How many times in your dreams have you heard the command, "It's not your time?"

Second, as also noted before, those who report a life review feel the impact of their words or deeds on their fellow human beings rather than themselves. That is hardly a component of our day-to-day personal memories. A mind greater than our own—perhaps a universe functioning as a quantum computer or a holographic universe or the mind of God—must provide a larger perspective.

---

[906] Michael Persinger, "Near-Death Experiences and Ecstasy: A Product of the Organization of the Human Brain," in *Mind Myths: Exploring Popular Assumptions About the Mind and Brain*, ed. Sergio Della Sala (Chichester, UK: Wiley, 1999): 85–99.

[907] Susan J. Blackmore and Tom S. Troscianko, "The Physiology of the Tunnel," *Journal of Near-Death Studies* 8 (September 1989):15–28, https://doi.org/10.1007/BF01076136.

[908] Susan J. Blackmore, "Out-of-Body Experience," in *The Skeptic Encyclopedia of Pseudoscience*, ed. Michael Shermer (Santa Barbara, CA: ABC-CLIO, 2002): 164–169.

[909] Robert Martone, "New Clues Found in Understanding Near-Death Experiences: Research Finds Parallels to Certain Psychoactive Drugs," *Scientific American*, September 10, 2019, https://www.scientificamerican.com/article/new-clues-found-in-understanding-near-death-experiences/.

Does Religion Matter?

Brain scans of Romanian orphans shocked the world with their findings: unloved infants had grossly underdeveloped brains. Normal brain development, it seems, requires an ample supply of love. As noted in chapter 24, children receiving adequate parental love develop brain connections "like a national highway system under construction."[910] Romanian orphans were deprived of a fundamental growth factor for the human brain: unconditional love.

The Christian faith asserts that connection with the component of the Holy Trinity known as the Holy Spirit confirms the power of love.[911] It has the power to change people and personalities. It modifies lifestyles. St. Paul summed up a list of what he called the fruit of the spirit in Galatians 5:22 (NIV). "love, joy, peace, patience, kindness, goodness, fidelity, gentleness, and self-control." As St. Paul assured the pagan philosophers in Athens, "For in him we live and move and are."[912] Wherever you are, that's where God is. The Holy Spirit is nothing less than the distributed and indwelling presence of God.

---

[910] Melissa Fay Greene, "30 Years Ago, Romania Deprived Thousands of Babies of Human Contact," *The Atlantic*, July/August 2020, https://www.theatlantic.com/magazine/archive/2020/07/can-an-unloved-child-learn-to-love/612253/.

[911] Gilles Emery, *The Trinity: An Introduction to Catholic Doctrine on the Triune God* (Washington DC: Catholic University of America Press, 2011).

[912] Acts 17:28 (NIV).

Asbury Awakening

A recent stirring of Gen Z students (individuals born between 1997 and 2012) at Asbury University in Wilmore, Kentucky, suggests that young people in their late teens and twenties continue to hunger for God's love.[913]

People born in the internet age often classify their religion as "none," reflecting their doubts about God. But after experiencing the brunt of a worldwide pandemic, Gen Z youth are facing a rising tide of anxiety and depression. A recent Centers for Disease Control and Prevention report noted that 44.2 percent of high school students describe "persistent feelings of sadness or hopelessness," with 19.9 percent "seriously considered attempting suicide."[914]

Perhaps that helps explain the reaction in an Asbury University chapel when a Missionary Alliance campus minister delivered a sermon focused on guilt, shame, anxiety, abuse, and distorted love. "Some of you guys have experienced radically poor love. Like evil love, selfish love, and I would say, today, we should not even give it the honor of calling it 'love.'" He continued: "Some of you guys have experienced that love in the church. Maybe it's not violent, maybe it's not molestation.... But it feels like someone has pulled a fast one on you.... This is not love."[915]

---

[913] "A Christian college in Kentucky has experienced a religious awakening," *The Economist*, February 23, 2023, https://www.economist.com/united-states/2023/02/23/a-christian-college-in-kentucky-has-experienced-a-religious-awakening.

[914] Sherry Everett Jones et al., "Mental Health, Suicidality, and Connectedness Among High School Students During the COVID-19 Pandemic—Adolescent Behaviors and Experiences Survey, United States, January–June 2021," *Morbidity and Mortality Weekly Report*, supplement, 71, no. 3 (April 2022):16–21.

[915] Terry Mattingly, "At Asbury, 2023 Revival Has Been 'Deja Vu All Over Again,'" Knoxville News Sentinel, March 2, 2023, https://www.knoxnews.com/story/entertainment/columnists/terry-mattingly/2023/03/02/terry-mattingly-at-asbury-revival-has-been-deja-vu-all-over-again/69950855007/.

In response to that sermon, Asbury students streamed to the altar to pray. Worshipers sang hymns and offered testimonies and public prayers. As a participant observed, "For seemingly no reason at first on Wednesday, February 8 [2023], the chapel service didn't end.... [A] young army of believers [rose] to claim Christianity...as their own, as a young generation and as a free generation, and that's why people cannot get enough."[916]

Ruth Graham, a religious writer for the *New York Times*, described the event: "Drawn by posts on TikTok and Instagram, plus old-fashioned word of mouth, Christians from across the country poured through a chapel on the campus of Asbury University to pray and sing until the wee hours of the morning, lining up hours before the doors opened and leaving only when volunteers closed the chapel at one a.m. to clean it for the next day."[917]

The quiet awakening at Asbury continued for the next eleven days. By its end, more than fifty thousand visitors had been drawn to Wilmore, Kentucky, representing more than two hundred academic institutions and multiple countries.[918]

One observer commented: "The wildness of these events is that they're actually un-wild. The atmosphere is serene, deep, and at times rather quiet.... It's like a veil is pulled back and students see Jesus for the first time—Jesus manifested in a new and powerful way."[919]

---

[916] Rebecca Paveley, "'Outpouring' at Asbury University," Church Times, February 24, 2024, https://www.churchtimes.co.uk/articles/2023/24-february/news/world/outpouring-at-asbury-university.

[917] Ruth Graham, "'Woodstock' for Christians: Revival Draws Thousands to Kentucky Town," *New York Times,* February 23, 2023, https://www.nytimes.com/2023/02/23/us/kentucky-revival-asbury-university.html.

[918] Fiona Morgan, "Citing Disruptions to School and Town, Asbury Authorities Move to End 13 Days of Revival," Religion News Service, February 20, 2023, https://religionnews.com/2023/02/20/citing-disruptions-to-school-and-town-asbury-authorities-move-to-end-11-days-of-revival/.

[919] Mattingly, "At Asbury, 2023 Revival."

An Asbury Seminary student described "a tangible sense of peace for a generation with unprecedented anxiety. A restorative sense of belonging for a generation amid an epidemic of loneliness. An authentic hope for a generation marked by depression.... A leadership emphasizing protective humility...for a generation deeply hurt by the abuse of religious power. A focus on participatory adoration for an age of digital distraction."[920]

An Asbury Seminary professor observed: "These students are hungry for a God who wants to change their lives. The true hunger for God is something students do not see in the culture around them. They are sick of trying to live without that."[921]

Orphans deprived of their mother's love have underdeveloped brains. Young people deprived of divine love suffer from anxiety, depression, and loneliness.

An article about the event interviewed a professor at Asbury Theological Seminary, who put the Asbury experience in perspective. "Revivals, he explained, begin with an awakening inside a Christian community—that's stage one. True revivals, throughout history, have led to evangelism, missions, and 'efforts for social justice' at the national and global levels."[922]

England in 1738

In 1738, religion and morality in Britain had collapsed "to a degree that was never known in any Christian country."[923] The royal court had persecuted the Puritans, prompting their departure to the col-

---

[920] Mattingly, "At Asbury, 2023 Revival."
[921] Mattingly, "At Asbury, 2023 Revival."
[922] Mattingly, "At Asbury, 2023 Revival."
[923] Diane Severance, "Evangelical Revival in England," Christianity.com, April 28, 2010, https://www.christianity.com/church/church-history/timeline/1701-1800/evangelical-revival-in-england-11630228.html.

onies in America. Deism, a belief system that denied the existence of a personal God, had overwhelmed the seminaries. The clergy of the Church of England had become corrupt, and their corruption spread throughout the culture.

At Oxford University, John Wesley formed a religious study group called the "Methodists" because of their stress on methodical study and devotion.[924] Also called the "Holy Club," they were notable for their steady communion services and for dedicating two days a week for fasting. Although Wesley never left the Church of England, his Methodism was destined to become a worldwide Protestant denomination.

The Methodist Revival

Wesley and his fellow Methodists began to declare the "glad tidings of salvation" to prisoners, poorhouses, and Church of England congregations.[925] Wesley embraced field-preaching as the most likely means of reaching the working class. Within the darkness of eighteenth century England, a Wesleyan revival broke out, and history credits that revival with the arrival of the Victorian age in the early nineteenth century.[926] The poor began to matter and "doing the right thing" became important. The Methodist Revival did nothing less than reclaim English homes and towns and the country from a pagan lifestyle, as individual redemption led to social regeneration.

---

[924]   Richard P. Heitzenrater, *Wesley and the People Called Methodists* (Nashville: Abingdon Press, 2013), 42.
[925]   Heitzenrater, *Wesley and the People Called Methodists*, 99.
[926]   Vishal Mangalwadi, *The Book That Changed Everything: The Bible's Amazing Impact on Our World* (Pasadena: Sought After Media Publications, 2019), 300.

The social effects of the revival were profound, extending to education, prison reform, hospital facilities, poor relief, temperance advocacy, and the abolition of slavery. A breath of new life swept through the church and the nation.

Mystical Experience

Mystical experiences have been the cornerstone of religious and spiritual practices for thousands of years.[927] From early Christian mysticism to Zen Buddhism, almost every religious path embraces the more mysterious aspects of reality. Mystical illumination, many observers suggest, may be the foundation of all religious experience.[928]

Mystical experience can occur during deep meditation,[929] following the administration of psychedelic drugs,[930] or at completely unexpected times, like during sport events or following interactions with nature or wild animals. Independent of the circumstances, mystical experiences share consistent features.[931]

1. They involve "dissolution of the ego." Neuroradiologists have identified the mechanism of this change in

---

[927] Aldous Huxley, *The Perennial Philosophy: An Interpretation of the Great Mystics, East and West* (New York: Harper Perennial, 2009).

[928] William James, *The Varieties of Religious Experience: A Study of Human Nature* (New York: Penguin, 1982).

[929] Tenzin Wangyal Rinpoche, *Awakening the Luminous Mind: Tibetan Meditation for Inner Peace and Joy* (New York: Penguin Random House, 2015).

[930] R. R. Griffiths et al., "Psilocybin can Occasion Mystical-Type Experiences having Substantial and Sustained Personal Meaning and Spiritual Significance," *Psychopharmacology* 187 (2006): 268–83, https://doi.org/10.1007/s00213-006-0457-5.

[931] W. N. Pahnke, "Psychedelic Drugs and Mystical Experience," *International Journal of Psychiatry in Clinical Practice* 5, no.4 (1969):149–62.

perspective. Psychedelics shut down a default mode network in the brain responsible for an overfocus on self.

2. During a mystical experience, space and time lose their meaning.

3. Experiencers encounter "cosmic consciousness" and a sense of oneness with the infinite.

4. The limitations of space and time give way to a sense of sacredness involving bliss, euphoria, ecstasy, and unconditional love.

5. Despite this radical change in perspective, mystical experiences feel more real than normal reality, as though they are revealing fundamental truths.

6. At the same time, mystical experiences are described as ineffable, beyond description by words.

Mystical experiences are short-lasting but immensely meaningful and profound. The net result: mystical experiences often change lives.

Johns Hopkins Center for Psychedelic and Consciousness Research

Although the government classified psychedelics as Schedule I drugs not fit for human testing, investigators at Johns Hopkins recognized their profound potential and petitioned the FDA to approve clinical trials using the active ingredient of magic mushrooms, psilocybin. A groundbreaking study was published in 2006.[932] During an initial trial of psilocybin, patients were evaluated serially with

---

[932] Griffiths et al., "Psilocybin can Occasion Mystical-Type Experiences."

well-recognized standard questionnaires used to evaluate the mystical experience. That questionnaire evaluated seven domains:[933]

1. Internal unity (pure awareness; a merging with ultimate reality)
2. External unity (unity of all things; all things are alive; all is one)
3. Transcendence of time and space
4. Ineffability and paradoxicality (claim of difficulty in describing the experience in words)
5. Sense of sacredness (awe)
6. Noetic quality or claim of intuitive knowledge of ultimate reality
7. Deeply felt positive mood (joy, peace, and love)

Two of Every Three Participants Had a "Complete" Mystical Experience

Based on criteria established before the study, twenty-two of the thirty-six volunteers treated with psilocybin in this study had a "complete" mystical experience.[934] Sixty-seven percent of the volunteers rated the experience with psilocybin to be either the single most meaningful experience of their life or among the top five most meaningful experiences. Thirty-three percent rated the psilocybin experience as being the single most spiritually significant experience of their life.

This groundbreaking, double-blind study demonstrated that psilocybin could reproduce a classic mystical experience. Participants commented that psilocybin treatment had substantial

---

[933] Pahnke, "Psychedelic Drugs and Mystical Experience,"
[934] Griffiths et al., "Psilocybin can Occasion Mystical-Type Experiences."

personal meaning and spiritual significance. Many participants believed they communicated with a *"conscious, benevolent, intelligent, sacred, eternal, and all-knowing power."*[935]

This encounter with the "ultimate reality" appeared to produce lasting benefits to mental health.

The Science of Psilocybin

A worldwide effort is ongoing to understand the action of psychedelics.

LSD, psilocybin, and DMT (the active ingredient in a plant-based brew called ayahuasca, used by indigenous peoples for thousands of years) produce altered states of consciousness. All three bind to the 5-hydroxytryptamine (serotonin) 2A receptor (5-HT2AR). In February 2023, investigators reported on a possible therapeutic mechanism.[936]

The 5-HT2AR receptor is expressed by multiple organelles within neurons in the brain. Serotonin binds to receptors on the cell membrane, but its chemical structure prevents it from penetrating the interior of the cell. The classic psychedelics readily penetrate the neuronal membrane, where they increase the growth of dendrites (the bushy extensions of the neuron) and "spine density." Spines are protrusions from the dendrites that house receptors for neurotransmitters and play an important role in the production of

---

[935] Roland R. Griffiths et al., "Survey of Subjective 'God Encounter Experiences': Comparisons Among Naturally Occurring Experiences and Those Occasioned by the Classic Psychedelics Psilocybin, LSD, Ayahuasca, or DMT," *PLoS One* 14, no. 4 (2019): 1–26, https://doi.org/10.1371/journal.pone.0214377.

[936] Maxemiliano V. Vargas et al., "Psychedelics Promote Neuroplasticity through the Activation of Intracellular 5-HT2A Receptors," *Science* 379, no. 6633 (February 2023): 700–706, https://doi.org/10.1126/science.adf0435.

synapses (connections with the axons of neighboring neurons). In other words, psychedelics promote neuroplasticity.

Treatment with psilocybin has shown promise for the treatment of depression,[937] nicotine addiction,[938] and relief of existential anxiety in cancer patients.[939] Treatment with psychedelics will gradually progress from small clinical trials to widespread application.

Skeptical materialists might dismiss mystical experiences as hallucinations. But like the reports of the near-death experience, descriptions of mystical experiences are remarkably consistent. Importantly, the benefit of psilocybin seems to require the induction of a mystical effect. Mother's love promotes infant brain development. Encountering divine love during a mystical experience has a remarkably similar effect.

---

[937] Alan K. Davis et al., "Effects of Psilocybin-Assisted Therapy on Major Depressive Disorder: A Randomized Clinical Trial," *JAMA Psychiatry* 78, no. 5 (November 2020): 481–489, https://doi.org/10.1001/jamapsychiatry.2020.3285.

[938] Matthew W. Johnson, Albert Garcia-Romeu, and Roland R. Griffiths, "Long-Term Follow Up of Psilocybin-Facilitated Smoking Cessation," *American Journal of Drug and Alcohol Abuse* 43, no. 1 (2017): 55–60, https://doi.org/10.3109/00952990.2016.1170135.

[939] Charles S. Grob et al., "Pilot Study of Psilocybin Treatment for Anxiety in Patients with Advanced-Stage Cancer," *Archives of General Psychiatry* 68, no. 1 (2011):71–78, https://doi.org/10.1001/archgenpsychiatry.2010.116.

Miracles

In our analysis of the relationship between the material brain and mind, we adopted a well-known position known as "substance dualism." Mind and matter are partners in creating human consciousness. But we adopted the Cartesian perspective, that the mind is the primordial substance.[940] Just as immaterial equations on an eternal mathematical landscape gave birth to the fields and forces of the universe and created matter in the first place, the mind will survive the death of our material body as a soul.

Richard Swinburne supports that conclusion. In *Are We Bodies or Souls?*, this emeritus professor of the Philosophy of the Christian Religion at the University of Oxford summarizes his perspective as follows: "Our bodies may be for the most part very complicated machines; yet we ourselves are not such machines, but essentially non-physical beings: souls who control bodies."[941] Substance dualism, he concludes, "holds that each of us living on earth consists of two distinct substances…but the part that makes us who we are is our soul…. By some miracle, we might continue to exist without

---

[940] Desmond M. Clarke, *Descartes, A Biography* (Cambridge: Cambridge University Press, 2006), 147.
[941] Richard Swinburne, *Are We Bodies or Souls?* (Oxford: Oxford University Press, 2019), 1.

our body, but no miracle would make it possible for us to exist without our soul."

C. S. Lewis

Most scientists consider a true miracle to reflect a violation of natural law—a parting of the Red Sea or the Incarnation. Materialists dismiss such events as physically impossible. But the great Christian apologist C. S. Lewis rejected their close-mindedness. What prevents the Maker of the Heavens and the Earth from further input? Lewis reasoned that if God "breathed" the universe into existence, He could breathe into that universe again.[942] The Architect, in other words, is free to supplement His plan.

Lewis focused on the momentous events of the Christian faith—the Virgin Birth, the Incarnation, and the Resurrection. Lewis cautioned, "God does not shake miracles into Nature at random as if from a pepper-caster. They come on great occasions: they are found at the great ganglions of history—not of political or social history, but of that spiritual history which cannot be fully known by men. If your own life does not happen to be near one of those great ganglions, how could you expect to see one? If we were heroic missionaries, apostles, or martyrs, it would be a different matter."[943]

Eric Metaxas

A more recent analysis sees miracles happening more frequently, not restricted to the "great ganglions" of history. Well-known author Eric Metaxas dreamed of ice fishing and catching a golden

---

[942] C. S. Lewis, *Miracles* (San Francisco: HarperOne, 2015), 69.
[943] Lewis, *Miracles*, 69.

fish. His dream led him back to his Christian faith and motivated him to draft a book devoted to its title, *Miracles*.[944]

In his definition, "A miracle is when something outside time and space enters time and space, whether to just wink at us or poke at us briefly, or to come in and dwell among us for three decades."

Miracles might include more than dreams. One might involve encountering the image of Jesus at the foot of your bed at a moment of emotional distress. Another might involve hearing the voice of God.

How do we reconcile these two models of miraculous events—miracles at the great ganglions of history or miracles encountered by ordinary people in everyday life?

On the one hand, we may be witnessing the work of God the Father, Maker of the Heavens and the Earth, attending to the needs of His son. On the other, we may see evidence of a triune God extending Himself through space and time through the power of the Holy Spirit.

The Holy Spirit

As per the Wikipedia article on the Holy Spirit, "In Judaism, the Holy Spirit…is the divine force, quality, and influence of God over the Universe and his creatures. In Nicene Christianity, the Holy Spirit is the third person of the Trinity[945]—a component of the triune God dwelling within each human being. People often describe the Holy Spirit as a presence or an "it," but theologians hold that

[944] Eric Metaxas, *Miracles: What They Are, Why They Happen, And How They Can Change Your Life* (London: Penguin, 2015).
[945] Wikipedia, s.v. "Holy Spirit," last modified July 19, 2014, 1:07, https://en.wikipedia.org/wiki/Holy_Spirit.

the Holy Spirit is a person, not a thing. The Holy Spirit is God. The Holy Spirit has thoughts and a will.[946]

During a mystical experience, individuals believe they encounter a "conscious, benevolent, intelligent, sacred, eternal, and all-knowing power."[947] Is this "all-knowing power" the Holy Spirit, the indwelling presence of God?

Whether described as "cosmic consciousness," the Holy Spirit, or "an all-knowing power," the miracle is that God is with us, whether in a moment of a mystical peak or in the quiet thoughts and prayers of everyday life.

Material Skepticism

David Hume was a Scottish Enlightenment philosopher, economist, and essayist in the mid-eighteenth century, best known today for his highly influential system of empiricism, naturalism, and skepticism.[948] This godfather of today's "scientism" forcefully denied the possibility of miracles. Hume, to put it simply, dismissed miracles as violations of the laws of nature.

What informed Hume's strongly held materialist worldview? Was he familiar with quantum mechanics? No, the behavior of particles at the quantum level was unknown. Was he aware of the curvature of space-time? No, and space-time remains a deep mystery today. Did he know about the Higgs field, the existence of particle physics, or the origin of life or DNA? The answer, of course, is no, which raises a fair

---

[946] Gilles Emery, *The Trinity: An Introduction to Catholic Doctrine on the Triune God* (Washington DC: Catholic University of America Press, 2011), 151.

[947] Roland R. Griffiths et al., "Survey of Subjective 'God Encounter Experiences': Comparisons Among Naturally Occurring Experiences and Those Occasioned by the Classic Psychedelics Psilocybin, LSD, Ayahuasca, or DMT," *PLoS One* 14, no. 4 (2019): 1–26, https://doi.org/10.1371/journal.pone.0214377.

[948] David Hume, "Of Miracles," in *Philosophical Essays Concerning Human Understanding* (London: A. Millar, 1748).

question: given his slim grasp of science, what was the foundation, other than unadulterated pride, for his strong opinions about miracles and nature?

Who are we to suggest that humankind will ever understand God's action in the world? But what beyond human pride warranted Hume's rejection of miracles over two centuries ago?

Violating the Laws of Nature

Hume and modern skeptics dismiss miracles as a violation of the laws of nature. St. Augustine provided a definition vastly closer to the truth: "Miracles are not a contradiction to nature. They are only a contradiction with what we know of nature."[949]

Recreation of the tomb of the crucified Jesus.

---

[949] Metaxas, *Miracles.*

An understanding of "sacred science" suggests that St. Augustine could have been bolder. Many aspects of science, albeit only partially understood, seem like miracles as well: the universe erupted from nothingness. A Higgs field filled the universe and encoded electrons and quarks with mass. The double helix of DNA delivers multiple codes for life.

Whether we agree with C. S. Lewis or Eric Metaxas, sacred science delivers a fundamental message: we are surrounded by miracles of the universe, life, and the mind every day!

The Mind–Brain Relationship: A Proposal

Supplement 19 reviewed the leading theories of consciousness, all focused on the interaction of material neurons with information, privileged brain centers, or electromagnetic fields. No theory has included the possibility that the elements of the mind are primordial, as independent, and essential as photons and quarks. The following hypothetical model is an attempt to predict how the units of the mind (quale) might interact with the physical structure of the material brain. As with all divine blueprints, the model has a mathematical foundation.

Neurons and their connections create more than a centillion (10^{303}) pieces of a jigsaw puzzle across the tissues of the brain. You might prefer to think of these jigsaw pieces as tiny maps. Each piece of the puzzle—each mental map, if you prefer—is associated with an individual component of qualia, the basic substance of the conscious experience and part and parcel of the "hard problem."

> **Key Definition: Neural Networks**
>
> *Billions of neurons (10^9) and trillions of synapses (10^{12}) build an inconceivable number of neural networks, mathematically mapping elements of mental content. Modern neuroscience is hard at work trying to identify those maps and their projections, hoping to define a complete human "connectome."*

As neuroscience develops the ability to identify these puzzle pieces or maps—the units of the "connectome"—they may reach a new understanding of the mental movie: is there a mathematical basis for quale assignment? Sequences map to genes and proteins. Will individual components of consciousness mathematically map to the incredibly complex puzzle of the brain?

The cumulative accretion of neural output from these maps will join with countless others at the speed of light to produce each frame of the mental movie. Those frames might gather on a special brain center, or alternatively, on the electromagnetic field overriding the brain.

As subjective experience calls each map into action, the brain can change. New connections may form. Memories may be created. The cumulative effect: a movie of our lives will premiere in a memory bank of the universe, yet to be defined. Is the universe a quantum computer? Is the cosmos a hologram storing the details of our lives?

In this model, "mind is what the brain does." But the "mind stuff" organized by the brain has an unimaginable and eternal source—nothing less than the mind of our Creator. We are both biological matter and self-conscious mind.

The mind uses the brain to create a unique creature, complete with body and soul. What happens to that mind at death? We should keep the faith: would the unimaginable Creator of the universe, the God who adorned His creation with the beauty of the lily, the hummingbird, and the nighttime sky not dwell in an even more beautiful home?

We should remember the words of Jesus: "In my Father's house there are many rooms." John 14:2 (NIV).

ACKNOWLEDGMENTS

When I was a newly minted cancer specialist from the National Cancer Institute and a new member of a small Sunday School class in Memphis, Tennessee, the group leader gave me an assignment. Would I lecture periodically on the relationship of science and faith? I remember my first lesson in 1982: Geologists had made the outrageous proposal that an asteroid smashed into Earth and ended the dinosaurs' reign. That theory proved true and gave me a foretaste of my science/faith career: expect surprising ideas. I am deeply grateful to the class's founder, Dick Ross (1924-2010), and the current class leader, Hank Shelton. Both encouraged the development of this manuscript.

Why would an oncologist cover a broad range of topics, from asteroids to quantum mechanics and DNA to theories of consciousness? First, the heterogeneity of cancer and the challenges of treatment humble every cancer specialist. We share a mission: to bring up-to-date agents to a treatment plan. We often use multiple agents in combination to maximize potential synergy.

That leaves us open to a broad range of scientific topics, and one issue demanded my attention. How do space-time, the Higgs mechanism, quantum mechanics, and the other impulses in the newborn universe impact the workings of the whole? Did God create the ultimate model of synergy through His powers of creation, from the universe and life to mind?

Few authors accomplish their intention without family support. How can I repay my wife Carole and daughters Lee and Meg, their husbands and children for their encouragement.

I am also eternally grateful for my parents, who believed life draws its meaning from its divine foundation. My surgeon father and servant-minded mother lived in the dark times of atomic war and Darwinian nihilism in the mid-twentieth century. But they kept their faith—in both our Creator and the sanctity of science. It gives me indescribable joy to see the best of science live comfortably with scripture today, confirming their hopeful outlook.

This book is also devoted to the patients and staff of the West Cancer Center and cancer patients around the world. Having lost my mother to breast cancer when I was twenty-one, I know your heartache firsthand. Our bodies are wondrously made, and research uncovers new treatments almost daily. But keep one truth in mind. God not only gave us sacred science. He also gave us promises of a sacred world to come.

I am also especially grateful to Madison Morris, who contributed numerous sketches distributed through the book.